Gender, Slavery and Law in Colonial India

INDRANI CHATTERJEE

AF740610

OXFORD
UNIVERSITY PRESS

OXFORD
UNIVERSITY PRESS

YMCA Library Building, Jai Singh Road, New Delhi 110001

Oxford University Press is a department of the University of Oxford. It furthers the University's objective of excellence in research, scholarship, and education by publishing worldwide in

Oxford New York
Auckland Bangkok Buenos Aires Cape Town Chennai
Dar es Salaam Delhi Hong Kong Istanbul Karachi Kolkata
Kuala Lumpur Madrid Melbourne Mexico City Mumbai Nairobi
Sao Paulo Shanghai Singapore Taipei Tokyo Toronto
and an associated company in Berlin

Oxford is a registered trademark of Oxford University Press
in the UK and in certain other countries

Published in India
By Oxford University Press, New Delhi

© Oxford University Press 1999

The moral rights of the author have been asserted
Database right Oxford University Press (maker)

First published by Oxford University Press 1999
Oxford India Paperbacks 2002

All rights reserved. No part of this publication may be reproduced, stored in a retrieval system, or transmitted, in any form or by any means, without the prior permission in writing of Oxford University Press, or as expressly permitted by law, or under terms agreed with the appropriate reprographics rights organization. Enquiries concerning reproduction outside the scope of the above should be sent to the Rights Department, Oxford University Press, at the address above

You must not circulate this book in any other binding or cover and you must impose this same condition on any acquirer

ISBN 0 19 565906 6

Printed in India at Pauls Press, New Delhi
and published by Manzar Khan, Oxford University Press
YMCA Library Building, Jai Singh Road, New Delhi 110001

Acknowledgements

The genealogy of this book is superficially traceable to a Ph.D dissertation written at SOAS, University of London, on a Felix Scholarship in 1996. The genealogies of the ideas in it should be rightfully traced to specific teachers at Delhi University, like Muhammad Amin, Sumit and Tanika Sarkar who initiated me in the art of questioning given truths. Other scholars who engaged with these ideas, and helped to refine them subsequently, include David Arnold, Peter Robb, Christopher Bayly, Michael Anderson, Richard Rathbone and Dorothy Stein in London; Bob Harms, Brian Fegan, K. Sivaramakrishnan, Liana Vardi, Kathryn and Henzpieter Znoj, and graduate students at the Programme of Agrarian Studies at Yale. Joseph Miller at Virginia, Paul Lovejoy at York, and Richard Hellie at Chicago enriched my stock of books and ideas with their generous gifts and correspondence on slavery in different hemispheres. The participants of the Oxford South Asian History Seminar, the Commonwealth Institute History Seminar, the British Records Association Seminar, and the graduate students of the Seminar on Feminist Theory at the London Institute of Education, and audiences in Delhi University and Calcutta University helped me to articulate some of the answers more forcefully than I would otherwise have done.

Sincere thanks to the staff of the British Library, especially the Reproductions Department and the Oriental and India Office Library, London, and of the Library at SOAS, of the National Archives, Delhi, the West Bengal State Archives and the National Library Calcutta and Nehru Memorial Library, Delhi. I would particularly like to thank Gautam Bhadra whose enthusiasm and archival aid went beyond any call of duty. I am grateful to the Maulana Abul Kalam Azad Institute of Asian Studies for a Senior Research Fellowship which gave me the time and leisure to tie up some of the loose threads. Fateh Ali Mirza spoke at great length about his family and put me in touch with Akhtar Hussain, the mutawalli of the Estate of Aman Ali Khan at Golkothi, Calcutta. To both the elders, a sincere thanks.

I thank the Lanning-Dinham household, and the Chisholm-Ellingham -Epstein household for incorporating me into their networks of support, and the Lokkas family for teaching me computerese and giving me a wonderful holiday in Greece, and friends Gita Sahgal, Kate Bennison, Kathy Prior, Ruth Vanita, Salim Kidwai, Supriya Guha, T. C. A. Raghavan and Tirthankar Roy, for extensive comments, enthusiastic encouragement and aid. I would also like to remember those who did not live long enough to read the conclusion of stories they had begun: my grandmothers, and my uncle Mridul. My parents, and my 'other' parents Samar and Jharna, my mother-in-law Mira Guha, and my aunt-in-law Amita Guha all stripped the gloss from terms of endearment in their recollections of a living past, and gave me the strength to understand the fragility of the 'familyscape'. To Sumit Guha, I owe all the Marathi and Portuguese sources and records cited in this work; were I to thank him for everything else, reams of paper would be insufficient. Finally, I would like to thank Rukun Advani at Oxford University Press, Delhi, and my editor Sikha Ghosh for easing the way of a first-time author. The flaws and errors of this book are, of course, mine alone.

Contents

Abbreviations

Actg	Acting.
AGG	Agent to the Governor-General.
BBOR	Bengal Board of Revenue.
BC	Board's Collection.
BCrJC	Bengal Criminal Judical Consultations.
BFP	Bengal Financial Proceedings.
BFoC	Bengal Foreign Consultations
BJC	Bengal Judicial Consultation and Proceedings.
BOR	Board of Revenue.
BPC/BPP	Bengal Political Consultation/Proceedings.
BPP	*Bengal Past and Present.*
BPubC	Bengal Public Consultations.
BRC	Bengal Revenue Consultations.
BS&M	Bengal Secret and Military Consultation.
BSBOR	Bengal Sudder Board of Revenue.
Capt.	Captain.
Ch.	Chief.
Collr.	Collector.
Commr.	Commissioner.
CPC	*Calendar of Persian Correspondence.*
Dept.	Department.
EPW	*Economic and Political Weekly.*
FPP	Foreign Political Proceedings.
GOB	Government of Bengal.
GOBy.	Government of Bombay.
GOI	Government of India.

Govt.	Government.
GPP	General Political Proceedings.
IESHR	*Indian Economic and Social History Review.*
IF&P	India Foreign and Political Proceedings
IHR	*Indian Historical Review.*
IHQ	*Indian Historical Quarterly.*
IJP	India Judicial Proceedings.
ILP	India Legislative Proceedings.
IPP	India Political Proceedings.
JAH	*Journal of African History.*
JAS	*Journal of Asian Studies.*
JASB	*Journal of the Asiatic Society of Bengal.*
Judcl.	Judicial.
Magt.	Magistrate.
Misc.	Miscellaneous.
MNLI	*Murshidabad Nizamut: Letters Issued.*
MNLR	*Murshidabad Nizamut: Letters Received.*
NAI	National Archives of India.
Offg.	Officiating.
OIOC	Oriental and India Office Collections.
Pers.	Persian.
PP	*Parliamentary Papers.*
Poll.	Political.
Sec.	Secretary.
SSRPD	*Selections from the Satara Rajas and the Peshwa's Diaries*
Suptdt.	Superintendent.
WBSA	West Bengal State Archives.

List of Nazims

1700–1727	Murshid Quli Khan, purchased as a child, trained in Persia, returned to India, served as viceroy to Mughal Emperor Aurangzeb.
1727–39	Shujā-ud-dīn, son-in-law (*dāmād*) of Murshid Quli Khan.
1739–40	Sarfarāz Khan, one of the sons of Shujā.
1740–56	Alivardi Khan, courtier and office-holder in Sarfarāz's government.
1756–57	Sirāj ud daula, grandson (son of daughter) of Alivardi Khan.
1757–60	Mīr Ja'far, brother-in-law of Alivardi Khan, and military paymaster (*bakshi*) under Siraj-ud-daula.
1760–63	Mīr Qāsim, son-in-law (*dāmād*) of Mīr Ja'far.
1765–66	Najm-ud-daula, son of Mīr Ja'far, mother Manni Begam who died in 1812.
1766–70	Saif-ud-daula, son of Mīr Ja'far, mother difficult to identify.
1770–93	Mūbarak-ud-daula I, son of Mīr Ja'far, mother Babbu Begam who died in 1809.
1793–1810	Babar Ali, Nāsir-ul-mulk, Mubarak-ud-daula II, son of Mubarak I, mother Faizunnisa, referred to as Wālida Begam.
1810–21	Zain-ud-dīn Alijah, son of Mubarak II, mother difficult to identify, but two principal consorts identified as Bahu Begam and cousin Amīrunnisa, referred to as Dūlhin Begam.

1821–24	Ahmad Ali Khan, Walajah, brother of Zain-ud-dīn, mother unidentified.
1824–38	Humayunjah, son of Walajah, mother Najībunnisā.
1838–80	Mubarak Ali Faridūnjah, son of Humayunjah, mother Sāhib Jān, also referred to as Ra'isunnisā Begam.
1880–95	Ali Kadr, son of Faridūnjah and Hasīna urf Mehrlakha Begam, succeeds to the title of Nawab of Murshidabad, but not to that of Nazim of Bengal.

1

Searching for Slaves in Indian History

Frederick Cooper once pointed out that colonialists in Africa and America refused to call a slave a slave, substituting words like serf, captive and dependent, thus avoiding having to confront slavery.[1] This occlusion of a widespread institution has perhaps nowhere been so complete as in the writing of South Asian history, where slavery has been displaced from narratives of power and statebuilding onto the interstices of social and economic history like that of jajmani.[2] Post-colonial historians have perhaps unconsciously contributed to this process by reiterating definitions and arguments that can be traced to twin sources: (a) the early British colonial enunciations on slavery both in India and in the Atlantic and (b) the historiography, and structure of, slave-based plantations in the northern Atlantic.[3] The questionnaire circulated by the Law Commissioners in 1834 understood slavery thus:

> in what they are employed and how they are worked? What species of produce are they employed in raising? Do they work in gangs, under a driver? for how many hours in the day?... Is the lash employed, and to both sexes?[4]

[1] Frederick Cooper, 'The Problem of Slavery in African Studies', *Journal of African History*, (henceforth *JAH*), 20, 1, 1979, pp. 103–25.

[2] C.A. Bayly, *Rulers, Townsmen and Bazaars: North Indian Society in the Age of British Expansion 1770–1870*, (Cambridge, 1983), p. 50, fn. 36. *See* Peter Mayer, 'Inventing Village Tradition: the Late 19th Century Origins of the North Indian "Jajmani" System', Modern Asian Studies, 27, 2, 1993, pp. 357–95.

[3] D. Banaji, *Slavery in British India* (Bombay, 1933); Amal K. Chattopadhyaya *Slavery in Bengal Presidency, 1772–1843* (London, 1977).

[4] Questions on Slavery in the East Indies, circulated by Commissioners for the Affairs of India, 15 March 1834, Slavery in India: Correspondence, *Parliamentary Papers* (henceforth *PP*), 1834, 44, no. 128, q.6., p.1.

English judges and magistrates in India responded by asserting that

> the slaves are not so systematically worked up, nor so cruelly whipped and punished as in the American slave-holding districts....[5]

Indian historians have therefore had to contend with standards of wageless and coerced labour set up in the historiography of the early modern Atlantic. Thus Irfan Habib argued that new crafts brought by new ruling groups in twelfth and thirteenth centuries required slaves trained in carding cotton, wheel-spinning and such skills. However, since slaves in productive work lacked 'the insecurity of subsistence, which drove and haunted free labour', they were soon replaced by the latter, and slave-markets declined in the fifteenth century.[6] Skilled slaves continued to be important in later centuries also,[7] but ideological blinkers prevented their incorporation into theories of work. Hence, denials of a 'slave mode of production' notwithstanding, a modernist constructed a typology of three kinds

[5] Extracts of Notes and Observations on Slavery, as existing in Bengal, Behar, and Benares, and the Ceded and Conquered Provinces by G. Myers, in Report from Indian Law Commissioners relating to Slavery in the East Indies, (*PP*) 1841, 28, no. 262, Appendix II, p. 281.

[6] Irfan Habib and Tapan Raychaudhuri (eds) *The Cambridge Economic History of India* (Delhi and Cambridge, 1984), pp. 92–3; and Irfan Habib, 'Slavery in the Delhi Sultanate, Thirteenth and Fourteenth Centuries—Evidence from Sufic Literature', *Indian Historical Review*, 15, 1–2, 1988–89, pp. 248–56. This is contrary to the arguments of Saleem Kidwai, 'Sultans, Eunuchs and Domestics: New Forms of Bondage in Medieval India' in Manjari Dingwaney and Utsa Patnaik (eds) *Chains of Servitude: Bondage and Slavery in India,* (Madras, 1985), pp. 76–96; J.E. Harris, *The African Presence in Asia: Consequences of the East African Slave Trade,* (Evanstone, 1971); and Muzaffar Alam, 'Trade, State Policy and Regional Change: Aspects of Mughal Uzbek Commercial Relations *c.* 1550–1750', *Journal of the Economic and Social History of the Orient* (hereafter *JESHO*), 37, 3, 1994, pp. 202–27.

[7] For trained female slave-performers, *see* Letter of Atmaram Rajaram in G.S. Sardesai (ed.), *Selections from the Peshwa Daftar* (henceforth *SPD*, Bombay, 1931), IV, p. 147. For skilled slave-cooks, seamstresses, and artisans, *see* Letter of Father John Cabral in *Travels of Fray Sebastien Manrique,* C.E. Luard (trans.), (Oxford, 1927), II, Appendix, p. 405. For eunuch-tutors, slave-writers and slave-commanders, *see* Ishwardas Nagar, *Futuhat-i-Alamgiri,* Tasneem Ahmad (trans. and ed.), (Delhi, 1978), pp. 8–9; S. Moinul Haq (ed.), *Khafi Khan's History of Alamgir* (Karachi, 1975), pp. 106, 432–3, J. Dowson (ed.), H.M. Elliott, *The History of India as Told by Its Own Historians* (London, 1877), VII, p. 150.

of slavery in colonial India. In one, called the domestic, she described the slaves of the indigenous elite and Europeans (in the metropolitan centres) as 'unproductive', despite being skilled.[8] The 'productive' labour such historians have on their minds is that which is extracted as surplus in the commoditized form of sugar, coffee, tobacco, cotton. Hence a second category—'agrestic' slavery,—becomes necessary. Yet, this typology takes little account of the existence of service/labour-obligations to both individuals and to the state in pre-colonial and colonial conditions, or the fact that an ideology of obligation structured all relationships in precolonial and early colonial India.[9] In some regions, like in precolonial Assam, male slaves were specifically exempt from military service or labour-obligations while non-slave males were not.[10]

Furthermore, the neat division of slavery into 'domestic' and 'agrestic' also deflected the violence borne by slaves,[11] from its position as the founding principle of intimacy in 'the family'. This was due to that 'conceptual apartheid' that had not only segregated women and the 'family' from the domains of men,[12] but had emptied one domain of the relations of domination while relocating them in another. Hence,

[8] Tanika Sarkar, 'Bondage in the Colonial Context' in Dingwaney and Patnaik (eds), *Chains of Servitude,* pp. 97–126. This typology is followed also in Ravi Ahuja, 'Labour Relations in an Early Colonial Context: Madras, 1750–1800', Paper at the Conference on Indian Labour History, Delhi, 1998.

[9] For a comparative perspective, *see* Anthony Reid (ed.) *Slavery, Bondage and Dependency in South-East Asia* (London and New York, 1983) and Tony Day, 'Ties That (Un)Bind: Families and States in Premodern Southeast Asia', *Journal of Asian Studies* (henceforth *JAS*), 55, 2, (1996), pp. 384–409.

[10] N.K. Barooah, *David Scott in North-East India 1802–1831* (New Delhi, 1970), p. 163, fn. 18 and Amalendu Guha, 'The Medieval Economy of Assam', *Cambridge Economic History of India,* I, p. 503.

[11] For the slave-boy killed by Lt. Arctander at Dinapore cantonment, see *PP,* 1828, pp. 61–72. The runaway-advertisements speak of the wounds of subjection as the identifiers of the slave, like the scars of a number of small burns on the back and legs, the chaffing of an iron ring on one leg, the cut little finger of a right hand in W.S. Seton Karr (ed.) *Selections from Calcutta Gazettes,* vol II, 1789–1797 (Calcutta, 1865) pp. 217–18 and 538–39. For a slave-*ayah* whose hair was shorn by her 'respectable' mistress, and another whose forehead was branded with the English word 'liar', *see East India Chronicle* (London) 6 August 1842, p. 174 and 9 May 1843, p. 106.

[12] Helen Bradford, 'Women, Gender and Colonialism: Rethinking the History of the British Cape Colony and its Frontier Zones, *c.* 1806–70', *JAH*, 37 (1996), pp. 351–70.

in coining the term 'domestic slavery', planters and their apologists in the eighteenth century implied that, by placing the slave and the mistress/master in close proximity, conditions of work and life had become favourable to the slave. Reinforcing the image of the plantation as a 'big happy family' provided multiple layers of moral defence for the slaveholder in both America and in England.[13]

Since the greater numbers of 'domestic' slaves are then seen (through an androcentric prism) to have produced no exchange value, and provision of crucial services and useful goods are imagined, a priori, as unproductive, the necessity of interrogating the basic postulates of 'class', genealogical pedigree, or the 'family' does not arise. Thus Marx and Engels, when schematizing the sequential development of the sexual division of labour in human societies, assumed a *prior* separation of monogamous wives from slave-concubines in the patriarchal family.[14] This has influenced historians to represent the female slave specifically as 'outside' the lineage and household, while the function of constituting the legitimate lineage was reserved to wives.[15] This is troubling not just because of the weight of subsequent historical evidence against such a divide, but because the inherently gendered and ageist conceptualization of 'productive' slave-labour erases various kinds of work performed by female and infant slaves *as* labour. This is evident in the historical treatment of the European households in colonial India.[16]

An implicit assumption undergirding this typology is that slavery was a phenomenon separate from the fabric of caste, community, lineage, religion. Here too historians appear to follow a well-trodden path. Macaulay's Law Commission assumed that slave-life was

[13] Margaret A. Burnham, 'An Impossible Marriage: Slave Law and Family Law'. *Law and Inequality*, 5, (1987), pp. 187–225.

[14] F. Engels 'Origin of the Family, Private Property and the State' in K. Marx *Selected Works* (Moscow 1977) III, pp. 238–9.

[15] *See* an instance of this model in Robin Blackburn, 'Defining Slavery—Its Special Features and Social Role' in Leonie J. Archer (ed.), *Slavery and Other Forms of Unfree Labour* (New York, Routledge, 1988), pp. 262–79. For a philosophical re-structuring of the priority of sexual subordination before the social contract, *see* Carole Pateman, *The Sexual Contract* (London, Polity Press, 1988).

[16] For a critique of the historiography, *see* my 'Colouring Subalternity: Slaves, Concubines and Social Orphans in Early Colonial India', *Subaltern Studies, X* (Delhi, forthcoming).

conducted in plantation-like segregation, and hence sought details of their presumably distinctive 'habits or morals' or 'marriage'. Hence, historians who rely upon these answers, discuss 'slave'-marriages as though slave-slave unions exhausted the range of ways in which slave sexual labours could be used by holders, as though slave-free cohabitational forms did not leak into and modify non-slave social forms. The implicit segregation of slavery from other aspects of socio-political life has impinged upon the conceptualization of the political economy of medieval India, forcefully highlighted by its marginality in the history of Mughal India.[17] However where the notion of segregated, coerced and unpaid labour has been most influential has been in the study of Indian caste.

Slavery and Caste

Dharma Kumar has recently stated that 'the term slavery does not accurately describe many forms of traditional bondage in (pre-colonial) India' which ranged 'from chattel slavery to debt-peonage'.[18] Using attributes such as restrictions on 'personal freedom', 'forced labor' and 'ownership' as criteria of status, Kumar claims that these also existed in the caste system. Evidently, the way historians conceptualize slavery immediately calls into question their conceptualization of caste itself.[19] If slavery is thought of as coerced and owned, permanently

[17] I.H. Qureshi, *The Administration of the Mughul Empire* (New Delhi, 1979); U.N. Day, *The Mughal Government 1556–1707* (New Delhi, 1969); M. Athar Ali, *Mughal Nobility under Aurungzeb* (Bombay, 1963), Stephen P. Blake, *Shahjanabad: The Sovereign City in Mughal India, 1639–1739* (Cambridge, 1991) and 'The Patrimonial - Bureaucratic Empire of the Mughals', *JAS*, 39, 1979, pp. 77–94. Richard M. Eaton, *The Rise of Islam and the Bengal Frontier, 1204–1760,* (Berkeley, 1993) discusses slaves in Sultanate Bengal only. An exception is J.F. Richards who outlines the centrality of the institution in 'Formulation of Imperial Authority under Akbar and Jahangir' and 'Norms of Comportment among Imperial Mughal Officers' in *Power, Administration and Finance in Mughal India* (Hampshire, 1993), pp. 252–89; also *The Mughal Empire* (Cambridge, 1992), p. 62.

[18] Dharma Kumar, 'Colonialism, Bondage and Caste in British India' in M. Klein (ed.), *Breaking the Chains: Slavery, Bondage, and Emancipation and Modern Africa and Asia* (Madison, 1993) pp. 112–30; *Colonialism, Property and the State* (Delhi, 1998), pp. 289–310; also *Land and Caste in South India* (Cambridge, 1965).

[19] Most scholars begin with the premise of the fixed nature of caste-ranking, for example *see* T.P. Vijaya, 'Aspects of Slavery in Coorg in the Nineteenth Century',

immobilized labour, then only the lowest *jatis* of the Indian caste system appear as 'slave-castes'.

Some of the attributes of caste were systematized by early English scholar-officials in India. Many of their ideas of identity (race) and status came from distinctly un-Indian locations. This is illustrated in the writing of a founder of Sanskrit studies—Henry Thomas Colebrooke. From the time of his arrival in India in 1782, he had entrepreneurial ambitions of becoming a plantation lord on the West Indian model; frustrated in South Asia he ultimately bought land in South Africa for this purpose.[20] In the Cape colony, he also witnessed slave sales without too many qualms.

In the light of Colebrooke's investment in the plantocratic ideal, his schematic arrangement of caste society in India coincided with the planter's preoccupation with male labour and its stabilization. In an unpublished preface to the *Digest of Hindu Law*, Colebrooke's admiration of the purportedly neat division of Indian society into 'slaves' and 'freemen' by the 'ancients' was re-aligned along occupational ranking that he interpreted as caste. Thus, he argued:

> Menial offices and mechanical labour were, in ancient times, executed by slaves, and deemed unworthy of freemen... it cannot appear strange that the class of *sudra* comprehended all servants and mechanics, whether emancipated or franchised, or descendants of emancipated persons. The freemen were denominated the twice-born... included, as was natural, the priest, the soldier, the merchant, and the husbandman... the Brahmana, Cshatriya and Vaishya;....[21]

The racially segregated Atlantic plantation was clearly the context in which Colebrooke categorized 'menial and mechanical' labour as specific to loss of citizenship, and approved of the exclusion of

---

Indica, 29, 1992, pp. 107–22. I have relied upon a radically different view of negotiated 'belonging' within a caste from Sumit Guha, 'An Indian Penal Regime: Maharashtra in the Eighteenth Century', *Past and Present*, 147, 1995, p. 116 and H. Fukuzawa *The Medieval Deccan: Peasants, Social Systems and States, Sixteenth to Eighteenth Centuries* (Delhi, 1991), pp. 91–113.

[20] T.E. Colebrooke, *The Life of Henry Thomas Colebrooke* (London, 1873), pp. 57, 316, 334.

[21] 'Heads of a Dissertation to be prefixed by way of Introduction to the Digest of Hindu Law, Civil and Criminal' in ibid., pp. 98–9.

descendants of slaves from such citizenship.[22] He had thus transposed a specific legal and social framework from the Atlantic on to the history of Indian caste.[23] The menialization of labour, and the idea of disabilities inherited in the blood were worked into early English representations of 'caste', displacing older European usages.[24]

Traces of this English remaking of caste—critical for the fashioning of early anthropology,[25] reappear in perspectives that posit a fixity of ritual status, and its separation from political-jural status. As Caplan's work on slavery in Nepal and India shows, the difficulty in separating 'slavery' from 'lower caste' status in India has become nearly

[22] We should not eliminate the possibility of the influence of Roman law on Colebrooke. However, the state of English eighteenth-century knowledge of Roman law is beyond the scope of this work. Historians of the present day insist that in Roman law, slaves manumitted according to one of the proper ways, became citizens. *See* a comparative study of the two legal systems in Alan Watson, *Slave Law in the Americas* (Athens and London, 1989).

[23] Gyan Prakash has urged that the separation of contaminating work from non-contaminating work in Narada's texts was appropriated by Colebrooke as the separation of the slave from the free, in 'Colonialism, Capitalism and the Discourse of Freedom' in Shahid Amin and Marcel van der Linden (eds) *International Review of Social History,* Supplement 4, 1996, pp. 9–25; *also see* Susan Bayly, '"Caste" and "Race" in the Colonial Ethnography of India' in P. Robb (ed.), *The Concept of Race in South Asia* (Delhi, 1995), pp. 165–218.

[24] *See* Gita Dharampal-Frick, 'Shifting Categories in the Discourse on Caste: Some Historical Observations' in Vasudha Dalmia and Heinrich von Stietencron (eds) *Representing Hinduism: the Construction of Religious Traditions and National Identity* (New Delhi and London, 1995), pp. 82–102. She argues that in the early decades of the sixteenth century the word 'casta' embraced several meanings, only one of which implied the 'purity of blood'; the terms more commonly found in the German reports were those employed for 'orders' 'grades' 'sections'. Entirely absent from the German reports are those features that characterize the English writing of the eighteenth century—the notion of a linear hierarchy and the unitariness of the system.

[25] The influence of Colebrooke's writing for Martin's edition of Buchanan's surveys, *The History, Topography and Antiquities of Eastern India* (London, 1835), for J. Wise, *Notes on the Races, Castes and Trades of Eastern Bengal* (London, 1883), and finally for Jogendranath Bhattacharya, *An Exposition of the Origins of the Hindu Caste System and the Bearing of the Sects Towards Each Other and Towards Other Religious Systems* (Calcutta, 1896, reprint Calcutta, 1973) is worth a separate investigation; *see also* Ronald Inden, *Imagining India* (Oxford, 1990).

insurmountable.[26] Caplan argued that 'domestic slavery' in South Asia was based upon a Dumontian dichotomy between power (political hierarchy) and status (ritual hierarchy).

Contrary to Caplan's assertion (based on Parliamentary Papers) that 'ritual rank was not fundamentally affected by whether the individual was a slave or not', most of the evidence available for India and Nepal indicates the instability of jati-identity of a slave and amends the invisibilization of gender in such an argument. For instance, of the Indian slaves (472) sent back from Nepal in the mid-nineteenth century, not even one was an adult male: the majority (312) were children, and the rest (160) were women.[27] Reporting the statements of some of these slaves, an English officer suggested that though they might have been originally from very low castes (*chamars, dosadhs, domes*) 'they have generally assumed to belong to higher castes such as *koormies, aheers, dhanooks, koeries* &ca. in order that they might be better treated.'[28] The difficulties of ascertaining original 'caste' ranking was well known in Bengal where two girls had 'been disposed off (sic) under feigned names... as Brahmini Girls'[29] while an Ahir woman was sold 'for Rupees 60 as a Rajpootin girl' in 1871.[30]

While demand for higher-ranking jatis may have been responsible for this 'passing', the evidence regarding pre-colonial regimes suggests that the ritual status of the person bought, or born of a slave, was corelated to political and other hierarchies. A narrative of the Rajas of Nadia in Bengal reported that 'the rajas purchased *sudra* boys and appointed them as their personal attendants; regardless of their jati they were proclaimed *kayastha* [of higher ranking in the varna arrangement]. Though these boys were initially degraded among the *kayastha sreni*, some of them have now become equal with the rest'.[31]

[26] Lionel Caplan, 'Power and Status in South Asian Slavery' in James L. Watson (ed.), *Asian and African Systems of Slavery* (Oxford, 1980), pp. 169–94.

[27] Statement of Slaves received from Nepal, Magt. Champaran to Commr. Patna Divn., 11 July 1867, WBSA, GPP, August 1867, no. 31.

[28] Resident at Nepal to Magt. Champaran, 10 June 1867, ibid., nos 11–12.

[29] Extract from Commr.'s Report, Judcl. Letter to Court of Directors, 27 June 1837, E/4/158.

[30] Crime Report, Patna Division, 1871, Appendix A, BJC, October 1871, no. 22.

[31] Karttikeya Chandra Ray, *Kshitish Vamsavali Charita,* (Calcutta, 1932), pp. 29–30.

This evidence destabilizes the orderly correspondence of jati-appropriate deployment of slaves, (*sudra* boys in physical proximity to the most Brahmanical of Bengal *rajas*).

Further evidence from Western and Central India confirms the fact of internal differentiation within each jati/caste, where slave-members, slave-born and non-slave members coexisted. An inscription of the eighteenth century illustrates this neatly.[32] Authorized by one claimant to the chiefship of Orchha, this inscription condemns the reigning chief, a descendant of Udet Singh, by describing the latter as having assimilated the children of his slave-women (*laundin ke jāide*) into his kin-group, and raised them to the chiefship. (In short, the inscription tries to ascribe slave-born status to the present chief). Then it proclaims that these people should not be assimilated into the Bundela jati from whose hands water may be drunk, nor should they be admitted into the ranks of those with whom one eats in assembly. Those who do admit these slave-born (*varanasankar, naichi panti*) to the ranks of their own jati, by eating with them and by marrying them, are in turn, severely cursed with all manner of degradation. As in the case of the Nadia rajas, this inscription, in trying to demean current practice, emphasizes the nature of a slave's jati-identity as derived from the interlocking political and social powers of their holders. The Peshwas, another stringently orthodox regime of the eighteenth century, could order the whole *got* of the *shimpis* (tailors) to admit Malhar, second husband of a widow and slave (*ghulam*) of Raghoji Kodhilkar of Saswad, into the jati. The grounds offered for this order are also instructive, since it says that it has been the practice in the jati to admit slaves into its ranks (*tumche yatimadhye sudamat ghulam jatit ghetat, yaise chālut āle āhe*).[33] The issue, it appears, was not the performance of labour corresponding with jati-status, but the *belonging within* a corporate group of the slave and slave-born. It was in this sense of 'belonging' that theoretically caste-free Muslim men and women when found in European households, spoke of themselves as having 'lost caste'; thus conceiving their experience as a loss of

[32] Hiralal, 'Sagar Ka Bundeli Shilalekh', *Nagaripracharini Patrika*, 8, (Samvat 1984) 1926, pp. 395–400.

[33] From the daily diary of the Peshwa, second half of the eighteenth century, cited in *Itihasa-sangraha*, 7, 1–3, 1915, p. 274.

belonging and status.[34] This belonging, marked by specific kinds of behaviours and ritual practices, was contentious for slave and slave-owners alike.

The ultimate arbiter of community membership was the regional hegemon such as the Peshwa in Maharashtra, or the various *dalapatis* and rajas in the case of Bengal. Thus, when a slave-born man and his accomplices claimed upward mobility *within* the Koshti (weavers) jati (from *kadu* or slave-born to *changli* or good), and their claims were contested by other Koshtis, the Peshwa ordered the segmentation of the kadu Koshtis from other Koshtis for purposes of marriage.[35] Nevertheless, it was nowhere denied that the slave-born were also Koshtis. This suggests that while the category ghulam or *laundi* was stable, ritual status or jati-identity in general, and that of slaves in particular, was open to negotiation and formalization in the late eighteenth century.

Despite the evidence, certain historians persevere with the association of low caste with slavery.[36] One of them has urged that the Peshwa's government enslaved predominantly its low-caste female subjects by the judicial device of implementing strict codes of sexual morality. However, the evidence from the same region and time reveals that *all* women, regardless of their jati-ranking, were vulnerable to enslavement for infringing moral codes, just as slave-women were vulnerable to 'outcasting' for the same reasons.[37] Whether or not a free woman was eventually enslaved depended on the strength of a woman's relationships with the other members of her *bhauband* (extended kin), their willingness and ability to pay her judicial fines, and subsidize her continued belonging in the jati.

[34] Evidence of Musst. Beebun, 1 February 1795 and Musst. Mairee, 13 February 1795, in Slavery in India: Correspondence, *PP,* 1828, pp. 67–8.

[35] G.C. Vad and K.B. Marathe (eds) *Selections from the Satara Rajas and the Peshwa's Diaries* (henceforth *SSRPD*), pt. 7, II, no. 763, pp. 337–9; for another instance of a slave-born son of a barber admitted into the got for ritual purposes, *see* R.V. Oturkar, *Peshwekaleen Samajik va Arthik Patravyavahar* (Pune, 1950), p. 89.

[36] Sharmila Rege, 'The Hegemonic Appropriation of Sexuality: The Case of the *Lavani* Performers of Maharashtra' in Patrica Uberoi (ed.), *Social Reform, Sexuality and the State* (New Delhi, 1996), pp. 23–38.

[37] *See* the case of the wife of a Prabhu who had behaved 'immorally' with an antyaja/untouchable man and had been enslaved in *SSRPD*, pt. 8, III, no. 1125, pp. 266–9; and two slave-women of Pune put out of caste for consorting with Frenchmen in ibid., no. 1103, pp. 251–2.

The Dialectic of Kinlessness and Kinship

It is obvious that conceptualizing slavery primarily as a labour-form, and therefore related to low caste-rank, is inadequate. Furthermore the age and gender profiles of such slaves do not commend themselves to such definitions. In the later years of the transatlantic trade, slaves imported into the American and West Indian colonies were predominantly male, both adult and children,[38] while the transferees described in South Asian evidence were mainly children, both male and female.[39] Surely agrarian tasks could not have been the primary work performed by these children and female slaves.

From the holder's perspective, infant and unskilled slaves were financially more attractive because larger capital outlays were required for the acquisition of an adult, presumably skilled and recalcitrant slave. All the Indian agents and *vakils* (pleaders and representatives authorized by litigants), many of them slave-holders, testified that children between five years and eight years of age were far cheaper to buy than young adult females or males.[40] It is also possible that a

[38] For recent statements of sex and age-composition of transatlantic cargoes, *see* David Eltis and David Richardson, 'West Africa and the Transatlantic Slave Trade: New Evidence of Long-Run Trends' and Philip Morgan, 'The Cultural Implications of the Atlantic Slave Trade: African Regional Origins, American Destinations and New World Developments', *Slavery and Abolition*, 18, 1, 1997, pp. 29–31 and 126–7 respectively.

[39] For African slaves between the ages of six and fifteen imported into India, *see* NAI, FPP, 19 September 1836, no. 10; 6 February 1837, no. 20; 24 January 1838, no. 11; NAI, FPP, A, May 1865, no. 90 and July 1883, nos 14–25. For slaves aged between two and twelve in intra-regional transfers, *see* Slavery in India: Correspondence (*PP*), 1828, 4, no. 125, pp. 11, 22–3; Report on Slavery in India (*PP*), 1841, Appendix II, p. 308; BPC, 10 May 1811, no. 109; BCrJ, 27 July 1837, no. 18; Panchanan Mandal, *Chithipatre Samajchitra* (Santiniketan, 1985), I, pp. 326–7 and Sivaratan Mitra *Types of Early Bengali Prose* (Calcutta, 1922) pp. 86–7. For comparisons with the ages of slaves in earlier centuries, *see* Richard C. Temple (ed.) *The Travels of Peter Mundy in Europe and Asia, 1608–1667* (London, 1914), II, pp. 88, 172; Shafaat Ahmed Khan (ed.), *John Marshall in India: Notes and Observations in Bengal, 1668–1672* (London, 1927), pp. 125, 137.

[40] Evidence of Raj Govind Sen, *PP*, 1841, Appendix I, p. 224. Tek Loll said 'children from six to eight sell for from 10 to 15 rupees' where adults cost between fifty and one hundred and twenty-five rupees, p. 225. The evidence of Dhurb Singh Das was that 'a boy of five or six years sells for one-fifth of the price of a young adult: the same of a girl', p. 231.

child offered a short-term speculative advantage as well as long-term investment, for she/he could be sold many times over. In 1791, one of the French ships transporting slaves from Calcutta to Pondicherry, carried on it seventeen girls between the ages of seven and seventeen.[41] Of these, 13-year-old Jannoo, 'slave to a person by the name of Bintoor', and the 10-year-old Aunchee 'slave to a Portuguese woman' were later found to have had at least two or three masters/mistresses in succession. The predominance of females in the internal hand-to-hand transfers may be partly explained by the value put upon reproductive abilities by indigenous informants in the eighteen-thirties, although price-differentials cannot be solely attributed to this.[42]

The predominance of children and females also forces us to rethink our definition of slavery. The demand for children was greater than the demand for adults, because the former

> recollect not after some time their parents and place of nativity and are more obedient than the grown up who are generally on the look-out for one early opportunity of running away.[43]

While the erasure of memory ensured the loyalty and dependence of the slave, for historians it means that establishing the primordial 'original' identity of such a person, either in terms of caste, creed, or lineage is impossible. First sales of children

> having happened when they were very young, they have lost all memory of their real parents, and do not even know their names, but believe themselves

[41] *PP,* 1828, pp. 31–3 and J. Motte and E. Maxwell, Police Office to Company's Attorneys, 2 June 1791, BFoC, 6 July 1791, no. 7. There are errors of copying and discrepancies between the two lists of the same ship: eight names of slave-girls and nine of the slave-boys listed in the *Parliamentary Papers* are absent from the list in the manuscript records of July, and the former list is less detailed than the latter.

[42] Evidence of Tek Loll, 28 December 1838, *PP,* 1841, Appendix I, p. 225. This witness said that the price of a young male was about a third less than the price of a young female. The fact that 'the girl may have children which will belong to her owner' only explains the differences between male and female, not differences of prices within a group of females. Nor can explanations resting on reproductive *ability* explain the fact that the highest prices were paid for eunuchs/castrated slaves. *See* evidence of Aga Kurbelai Mahomed, *ibid.,* p. 240.

[43] Resident, Gwalior to Sec. to the Gov.-Gen. at Shimla, 30 July 1832, BC F/4/1467/57728. Unless otherwise stated, all eighteenth and nineteenth century records are from OIOC, London.

to be children of their purchasers, or if not this, neither know nor care who their parents are... no clue would be obtained to their origin.[44]

A more nuanced understanding of slavery is essential in Indian history where slaves originated outside as well as within the sub-continent. Hence Orlando Patterson's characterization of the slave as the person who is held in a state of natal alienation, dishonour and violence, is useful as a point of entry. Accordingly, the most distinctive attribute of the slave's powerlessness originated conceptually as a substitute for death, archetypically for death in war, but equally frequently as commutation of capital punishment, or death from starvation and exposure. This conditionally pardoned 'life' was then to be lived in terms of permanent natal alienation, having no socially recognized existence outside of the saviour/master, and such a person was denied all claims on, and obligations to parents and living relations, as well as more remote ancestors and descendants.[45]

Patterson's formulation, however, must be qualified by two considerations. The first is that the idea of slavery as an alternative to death was the moral legitimation provided by slave-masters and acquirers. William Jones spoke of his own slaves as those 'whom I rescued from death or misery', as did the Governor-General, John Shore.[46] This legitimation was then extended to stave off abolition because, it was urged, the sellers were 'saving... their children's lives, by interesting in their preservation persons able to provide nourishment for them'.[47] Early English semantics of charity, of the rescue of children

[44] J. Beames, Magt. Champaran to Commr. Patna Divn., 11 July 1867, WBSA, GPP, August 1867, no. 30.

[45] Orlando Patterson, *Slavery and Social Death* (Massachusetts, 1982), pp. 5–13 and passim; for a review *see* Peter Kolchin, 'Some Recent Works on Slavery Outside the United States: An American Perspective', *Comparative Studies in Society and History* (henceforth *CSSH*), 28, 4, 1986, pp. 767–77.

[46] For the former, *see* S.N. Mukherjee, *Sir William Jones: A study in Eighteenth Century British Attitudes to India* (Cambridge, 1968), p. 134 and 'A Note on Sir William Jones and the Slave Trade in Bengal', *BPP*, 82, 154, 1963, pp. 106–11. For the latter, *see* F.J. Shore *Memoirs and Life of Lord Teignmouth* (London, 1837), I, pp. 156–7.

[47] H.T. Colebrooke, Minute of 1812, BC F/4/1234/40338. This attribution of charity and humanity persisted in the rhetoric of the officials who opposed abolition as late as the 1840s. Examples are the minutes of H.T. Prinsep, A. Amos in BC F/4/1947/84542.

from death, went alongside fairly arbitrary ascriptions of affect, or lack of it, to Indian kinship relations and terminologies. Secondly, in Patterson's formulation, the correspondence of affective with biological kinship (natal ties) is presumed to be already established. It was similarly assumed by English observers in India when they characterized slavery as a state in which 'the tender ties of parental, filial, and fraternal affections are dissolved'.[48]

In fact, early ethnographers like Hamilton Buchanan, Montgomery Martin and James Taylor, moved by the idea of the labouring slave, compiled vernacular terms denoting slave-status but underestimated the force of contemporary indigenous terms of kinship. Hence Martin characterized as 'not slaves' the children purchased and reared as *palokbeta* (sons by nourishment) in northern Bengal, even as he noted how he had been deceived by the use of multiple, simultaneous and different meanings of terms like *nufur, bahaiya, sudin, sehana* in Purnea.[49] Similarly, Taylor supplied the words current in eastern Bengal—ghulam, *bhandari* (for males) and *dasi, bāndi* (for females)[50] but was relatively silent on the content of kinship terms also operative in the cultures he described. Two centuries later, Amitava Ghosh[51] selected for his discussion of untranslateability the terms which fall outside the realm of kinship. Noting that the eighteenth century official, Buchanan Hamilton did not provide the Tulu or Kannada terms for 'slave', Ghosh compensates for this by supplying the Kannada word *tottu*, 'an ambiguous term, used for servants and hired workers as well as "slaves"'. The absolution that metaphors of kinship and affinity enjoy from Ghosh's searching analysis then endow the marriage of Ben Yiju and the Nair girl, pawned by her brother, with great affect; the author imagines the return of the slave-holder to his 'wife' in the conclusion. Similarly, English officialdom's notions of kinship simultaneously over-inscribed affective relations onto commercial ones,

[48] James Forbes, *Oriental Memoirs: Selected and Abridged from a Series of Familiar Letters Written During Seventeen Years in Residence in India* (London, 1813), III, p. 168.

[49] R.M. Martin *History, Topography, Antiquities*, III, pp. 121–2, 497.

[50] James Taylor, *A Sketch of the Topography and Statistics of Dacca*, (Calcutta, 1840), p. 320.

[51] Amitava Ghosh, 'The Slave of Ms. H.6' in Partha Chatterjee and Gyanendra Pandey (eds), *Subaltern Studies, VII*, (Delhi, 1993), pp. 159–220.

and erased the obligations of deference and labour owed by such purchased/captured persons. An official, discussing the sales and purchases of children in the Muzzaffarpur district of Bihar, confidently asserted that:

> After purchase the condition of these children is often far better than it was before, and though they are considered the property of the master, are not looked upon in the light of a slave,... often acquire property, and in some cases, even attain to great opulence.
>
> The child when purchased is fed and clothed by its master, and when it gets to sufficient age to be of any help, is given a small monthly allowance as well...
>
> To call either one custom or the other slavery would be giving a very wrong impression.[52]

Our inability to interrogate English-authored ascriptions of affect to 'family' relations constituted through purchase and pawnage leads to a replication of the ideology of paternalism forwarded by slave-holders themselves. This is illustrated by the contradictory pulls in recent and sophisticated studies like that of Gyan Prakash.[53]

Prakash began his work by asserting that official colonial ideas of slave-use led to the representation of the 'long-term ties' between the *kamias* (agricultural labourers) and their *maliks* (masters) in South Bihar as slavery, when it was not. Though I partly favour such a formulation, caution is necessary because scores of colonial officials painted Indian slavery in these distinctive paternalist colours.[54] Prakash argues that the key moment where this long-term tie is made visible is when the landlord grants the loan needed for the marriage of the labourer. This serves the labour needs of the landlord by reproducing both labour and dependence, and it subordinates the Kamia woman both to her spouse and the malik (thus reproducing the patriarchal

[52] Offg. Magt., Tirhut to Commr. Patna Divn., 30 September 1868, BJC, March 1869, no. 9.

[53] Gyan Prakash, *Bonded Histories: Genealogies of Labour Servitude in Colonial India* (Cambridge, 1990) and 'Terms of Servitude: The Colonial Discourse on Slavery and Bondage in India', in Klein (ed.), *Breaking the Chains*, pp. 131–49.

[54] For a fuller discussion, *see* Peter Robb (ed.), *Dalit Movements and the Meanings of Labour in India* (Delhi, 1993), Introduction, esp. pp. 26–43.

family among the kamias).[55] Hence patrilineal descent-reckoning bleaches the productive labour of Kamia women out of the legal bonds executed between men (kamia and malik) and drives out their reproductive labour from the oral traditions and Bhuinya genealogies. Philosophically, Prakash's argument is unexceptionable in the present. However, is that the story of the past as well? Furthermore, there is a noticeable absence of the 'father' from the origin-stories Prakash cites. The subsequent retrieval of paternity in the realm of overwhelmingly male ancestor-spirits suggests that Prakash is perhaps too quick to replicate the Kamia 'family' as a 'patriarchal' one. The malik's claims upon the labour and persons of the kamia women, when read alongside the kinship terms of the Bhuinya songs (pp. 51–2), further suggest that genealogies of patrilineal descent (the core of the 'patriarchal family') are constituted by the malik's power, rather than by the kamia male's procreative sexuality. The issue that has perplexed students of the plantation economies[56]—namely, the formation of slave-families—has conversely, been too early resolved by Prakash. Did the shared paternalism of the kamia male and malik in the matter of the marriage-payments, in fact, represent a tiny victory in acquiring the privilege of having a 'family' at all? Did the male kamia have the same access to non-kamia females as the male malik? Did polygyny and hypergyny affect male and female kamias differently, as well as male malik and kamia differently? Indeed, how did paternalism and slave-sociality relate to each other, if, as suggested by the origin-stories, kinship ties were tense, fragile, even malevolent?

Both Ghosh and Prakash raise significant questions for historians of kinship and community. In conditions in which some kin, like parents and uncles alienated, sold, or mortgaged their 'natal kin', the implications for the twin histories of wealth-creation and kinship-

[55] Gyan Prakash, 'Introduction: The History and Historiography of Rural Labourers in Colonial India' in Prakash (ed.) *The World of the Rural Labourer in Colonial India* (Delhi, 1994).

[56] *See* Herbert G. Gutman, *The Black Family in Slavery and Freedom, 1750–1925* (New York, 1976); Verena Stolcke, 'The Slavery Period and Its Influence on Household Structure and the Family: Jamaica, Cuba and Brazil' in Elza Berquo and Peter Xenos (eds), *Family Systems and Cultural Change* (Oxford, 1992), pp. 125–43; Barry W. Higman, *Slave Population and Economy in Jamaica, 1807–1834* (Cambridge, 1976); Barbara Bush, *Slave Women in Caribbean Society, 1650–1838* (Indiana, 1990).

formation need to be reassessed.[57] It is imperative for historians not to talk of terms of service alone but of terms of endearment.

Metaphors of Kinship: A Slave by Some Other Name

Analysing kinship in historical time is useful for epistemeological and functional reasons. Part of the problem is discursive: like other discourses fashioned in the Atlantic colonies, 'slavery' carried enormous moral and political charges. Uncomfortable with these, Claude Martin described the five young girls in his household,

> not as we term slaves, though paid a consideration for, but the sum I paid was a present to the relations, that I might have had a right on them as not to be claimed by anybody...
>
> ... I acquired them every one when they were young and children and took care of them by my acquiring them they have lost their fathers and Relations....[58]

William Jones' discomfort at the African trade (which his biographers extoll) did not change his assessment of slavery as an institution, which 'properly explained can produce little mischief'.[59] Jones' hatred of the word was important for the linguistic shift in the field both in his time and subsequently. He preferred the term 'servant' just as Colebrooke preferred the term 'apprentice'.

English slave-holders were not singular. Kinship terms supplement references to legal status in the case of slaves under the Sultanate and the Mughals also. Thus Aibak 'gave the title of son to Iltutmish and honoured him by keeping near his own person', and Muhammad Ghori described his Turkish male slaves as 'sons... they will inherit my lands and continue the Khutba in my name when I am dead and gone'.[60] In 1582, Akbar is said to have announced a change of nomenclature

[57] For an economic anthropology of such transfers of kin, *see* Charles Piot, 'Of Slaves and the Gift: Kabre Sale of Kin During the Era of the Slave Trade', *JAH*, 37, 1, 1996, pp. 31–49; for a theoretical reformulation, *see* Jane Guyer, 'Wealth in People, Wealth in Things–Introduction', *JAH*, 36, 1995, pp. 83–90.

[58] Will of Claude Martin, L/AG/34/29/12, 1800, pp. 121–61.

[59] Cited in Mukherjee *Sir William Jones*, p. 134.

[60] Dharam Pal, 'The Influence of the Slaves in the Muslim Administration of India', *Islamic Culture*, 18, 1944, pp. 409–17 (reprint, London, 1971).

for the Imperial slaves, who were to be henceforth styled *chelas.*[61] The Bangash nawabs of Farrukhabad referred to their four thousand male slaves as 'sons of the state' (*tifl-i-sarkar*)[62] just like the Marathas referred to enslaved women as 'daughters of the state' (*rajbeti*).[63] Holders and users attempted to soften the starkness of alienation experienced by slaves by changing nomenclature and suggesting another set of social relationships. Terms like *beti,*[64] *dharambap,*[65] *kanyapatri*[66] repeatedly crop up in the vernacular sources. Slaves and slave-descendants too normally refused to identify themselves as such.[67] Thus one English Collector recalled how the son of an old man from Sylhet 'would not allow that his father was actually the slave of the high-priest, but styled him his salt-eater, or dependent'.[68]

In social formations where the idiom of kinship provided a paradigm for the representation of wealth and power, the metaphoric

[61] Badauni attributed the use of the latter term to the influence of the mendicants called *jogis,* hinting at a form of borrowing nomenclatures from *within* the same cultural milieu, referring to the existence of child-oblates devoted to a spiritual corporate body. *See* Irfan Habib, 'Akbar's Social Views: A Study of their Evolution', Papers of the Indian History Congress, 53rd Session, 1992–93, p. 223. This practice of mendicant-worriors purchasing males, training them for diplomacy, war and commerce appears to have continued into the late eighteenth century, *see* Ananda Bhattacharya, 'Sanyasi and Fakir Uprising in Bengal in the Second Half of the Eighteenth Century', unpublished PhD. Diss., University of Jadavpur 1992, p. 183. For suggestive analogies, *see* John Boswell, *The Kindness of Strangers: The Abandonment of Children in Western Europe from Late Antiquity to the Renaissance* (New York, 1988), pp. 296–321.

[62] William Irvine, 'Bangash Nawabs of Farrukhabad', *JASB,* 47, 1, 1878, pp. 279–345 and *The Army of the Indian Moghuls: Its Organisation and Administration* (London, 1903), p. 103.

[63] Information collected by Vinayak Rao Aurangabadkar for Richard Jenkins, Resident at Nagpur, *c.* 1813, *see* OIOC, Mss. Marathi D. 31, folio 100.

[64] Slave-hire deed in Mitra *Types of Early Bengali Prose,* no. 2, p. 86; for similar terms in Assam, *see* Report on Slavery (*PP*), 1841, Appendix VI, no. 5, p. 404.

[65] Dinesh Chandra Sen *Bengali Prose Style: 1800–1857* (Calcutta, 1921), pp. 17–19.

[66] R.M.Martin *History, Topography and Antiquities,* III, p. 555.

[67] *See* National Library, *Kitab-i-Qissa Tahmas-i-Miskin,* J.N. Sarkar Mss. 174, or P. Setu Madhava Rao abridged and trans. *Tahmas Nama: The Autobiography of a Slave* (Bombay, 1967) for the self-description as *tifl* (boy/son).

[68] Extract from the *Lives of the Lindsays* in W.K. Firminger (ed.) *The Sylhet District Records, I,* (Shillong, 1913), p. 130.

representation of impoverishment and powerlessness was that of 'kinlessness'.[69] The idiomatic representation of powerlessness-as-kinlessness was significant because assimilation with the host society or lineage or household was an important demand put upon slaves. At the same time slaves, like others, were themselves measures of the power and dignity of others. This is why the acquisition of *female* and *infant* slaves was so characteristic of South Asian households.

Perhaps the reluctance of the Indian historian of the 'family' to name slave-recruitment as a principal aspect of nineteenth-century households and kin-structures has been due to the fact that the issues and problems have been conceptualized in the light of a persuasive and intensively researched historiography of family, kinship and household in the West.[70] However, while the historicity of the modern European state-system and the family can be established, it cannot be emulated for a history of the state and the family in precolonial India.[71] For one, the Western European family grew first with slaves in it, and alongside slavery through the entire early modern period. Unlike the debate, set forward by Eric Williams,[72] and participated in by other historians,[73] on the contribution of slavery to the transition to

[69] *See* my 'Testing the Local Against the Colonial', *History Workshop Journal*, 44, 1997, pp. 215–24.

[70] Lawrence Stone, *The Family, Sex and Marriage in England, 1500–1800* (New York, 1979); Peter Laslett (ed.), *Household and Family in Past Time* (Cambridge, 1972); Eli Zaretsky, *Capitalism, the Family and Personal Life* (New York, 1976); Jean Louis Flandrin, *Families in Former Times*, Richard Southern (trans.), (Cambridge, 1979); Stephanie Coontz, *The Social Origins of Private Life: A History of American Families 1600–1900* (London, 1988); Leonore Davidoff and Catherine Hall, *Family Fortunes: Men and Women of the English Middle Class, 1780–1850* (London and Chicago, 1987) and Catherine Hall *White, Male and Middle-Class: Explorations in Feminism and History* (Oxford, 1992) among others.

[71] For a parallel critique of Eurocentric models for the study of Africa, *see* Steven Feierman, 'Africa in History: The End of Universal Narratives,' in Gyan Prakash (ed.), *After Colonialism: Imperial Histories and Post-Colonial Displacements* (Princeton, 1995), pp. 40–65.

[72] Eric Williams, *Capitalism and Slavery*, (London, 1944).

[73] Amongst a vast and sophisticated literature, *see* E.D. Genovese and E. Fox-Genovese, *Fruits of Merchant Capital: Slavery and Bourgeois Property in the Rise and Expansion of Capitalism* (New York, 1983); Robert W. Fogel and Stanley L. Engerman *Time on the Cross:* vol. I, *The Economics of American Negro Slavery*, and vol. II, *Evidence and Methods* (Boston, 1974).

capitalism in the Western hemisphere, there has been virtually no work on the 'expulsion' of slaves from the Western 'family', where they were firmly lodged in Locke's time. The centrality of slavery to 'family' and household has been particularly noticed by students of Western colonial settlements such as that of South Africa and India,[74] but has not generated debate in the case of the family in Western Europe.

Secondly, the separation of the 'state' from the 'family' in Western Europe was fundamental to the way an ideology of domestic/public or private/public was worked out. As feminist historians have pointed out, though 'public' came to represent the state, marked by power, autonomy and activities of males in Western European thought and social organization, it did not follow that wherever these dichotomous representations existed the attribution of values was identical. Furthermore, the separation of the 'state' and the 'family' was by no means a foregone matter in the case of eighteenth-century India. This conjuncture is even more significant for historians of slavery because of the decrees of exclusion passed by the Court of Directors.

In its instructions to the Law Commissioners, the Court recommended that certain social forms be excluded from the ambit of slavery.[75] What some of these forms were can be inferred from the debates in Parliament; in the Commons, the central issue seemed to be the 'harems and zenanas', an intrusion into which would excite rebellion.[76] Macaulay, Secretary of the India Board, even reassured the House that since 'due regard shall be paid to the laws of marriage; those who live in the zenanas may be considered as coming under this class, the connection in this case is a quasi-marriage'. Why marriage? English and later American abolitionists, inspired by Evangelism, criticized planters for preventing slaves from organising gender in the manner that both sides in the debate assumed was best. Slavery

[74] Robert C.-H. Shell, *Children of Bondage: A Social History of the Slave Society at the Cape of Good Hope, 1652–1838* (Hanover, 1988); *also* Christopher Hawes *Poor Relations: The Making of an Eurasian Community in British India, 1773–1833* (London, 1996).

[75] Extract Public Letter to India, 10 December,1834, in Slavery in India: Correspondence (*PP*), 1834, 44, no. 128, p. 70.

[76] Hansard, *Parliamentary Debates*, 3rd. series, (London, 1833), XIX, pp. 532, 798–9.

endangered family relations because husbands and wives could always be sold and separated, discouraging slaves from marrying, slave-parents from caring for their children, and ultimately resulted in promiscuity and demoralization of slaves. As Paton notes, participants on all sides of the debate shared the belief that fornication defined sexual sin. Thus marriage, not consent, was the key factor in determining the morality of an interaction.[77] In the face of an abolitionist agitation, the simple reduction of the complex and different grades of slavery into 'marriage' relations by the Law Commission absolved the Company of any responsibility to legally end slave-concubinage in the English officers and soldiers' households. On the other hand, the acute awareness of slave-concubinage within indigenous ruling houses enabled administrators to manipulate political succession and finances. Furthermore, this manipulation, as Lovejoy and Hagendorn argue, entailed curtailing the web of claims and rights accorded to slaves and slave-born in Islamic and indigenous legal structures.[78] In a similar vein, Auckland, the Governor-General, in 1841, recommended that sales of children may be prohibited but parenthesized 'the national custom of adoption' as one to be guarded.[79] However, when there were states whose revenues were to be appropriated, or whose territories the Company needed to annex for commercial or political advantage, such 'adoptions' of 'foundlings' provided the perfect justification for the overthrow of a dynasty or ruler. Instead of resisting the fiat of

[77] Diana Paton, 'Decency, Dependence and the Lash: Gender and the British Debate over Slave Emancipation, 1830–34', *Slavery and Abolition*, 17, 3, 1996, p. 169.

[78] Paul Lovejoy and Jan S. Hagendorn, *Slow Death for Slavery: the Course of Abolition in Northern Nigeria, 1897-1936* (Cambridge, 1993), pp. 111–24; *also* Paul Lovejoy, 'Concubinage and the Status of Women Slaves in Early Colonial Northern Nigeria', *JAH*, 29, 1988, pp. 245–66.

[79] Minute of the Governor-General, 6 May 1841, (*PP*) p. 478. In a private letter to Hobhouse, 1 April 1839, however, Auckland had accepted that various officers and magistrates 'only look to praedial slavery.... The abominations which grow out of the maintenance of the Jenana and the troops of dancing girls are of another character and for the most part illegal, but from national customs difficult to check....' in British Library, Broughton Papers, Add. Mss. 36473, folio 447. In the same collection is a letter from W.H. Macnaghten, 8 April 1837, admitting that the condition of female slaves was well known to officials, but '...who is to explore the recesses of the haram?', folio 110.

exclusion, all published work on Indian slavery has followed it, refusing to detail how the practices of concubinage/secondary marriages, or of adoption worked as a fundamental part of slave social formations. This has resulted in displacing issues of reproductive politics from the context of a slave-holding society and their relocation within other kinds of historical narratives, like those of the contest between imperialism and nationalism, or micro-level studies in social anthropology.[80]

Another sound reason for the analysis of metaphors and structures of kinship is that it sensitizes the historian to both market and non-market transactions in slaves. When hard pressed to pay fines or debts or redeem mortgages, many holders transferred such wealth-in-people to creditors and revenue-collectors. The transactions described by travellers, and the English officials in the *Parliamentary Papers,* as the sales of 'sons' and 'daughters' were transactions in children who were already slaves. As the Agent at Bareli discovered in 1811, the infant-slaves at the towns of Najibabad and Ujjaini—'established marts where these children are collected in hundreds'—were sold not by the parents of the children as had been imagined by all the officials, but by the middlemen slave dealers.[81] Crises like famines, and floods, were special moments associated with these sales. Since transfer during distress was associated with the inability to maintain one's dependents, and ushered the need for liquidating one's holdings,[82] the market was one where acquisition was associated with 'respectability' while sale,

[80] *See, for example,* K. Sangari and S. Vaid (eds), *Recasting Women: Essays in Colonial History* (New Delhi, 1989); M. Borthwick, *The Changing Role of Women in Bengal, 1849–1905* (Princeton, 1984); Malavika Karlekar, *Voices from Within: Early Personal Narratives* (Delhi, 1991); Prem Chowdhry, *The Veiled Women: Shifting Gender Equations in Rural Haryana 1880–1990* (Delhi, 1994). For micro-level studies of kinship and marriage, *see* Lina M. Fruzetti, *The Gift of a Virgin: Women, Marriage and Ritual in a Bengali Society* (New Delhi, 1990); Imtiaz Ahmed (ed.), *Family, Kinship and Marriage Among Muslims in India* (New Delhi, 1976); Jack Goody *The Oriental, the Ancient and the Primitive* (Cambridge, 1990); Patricia Uberoi (ed.), *Family Kinship and Marriage in India* (Delhi, 1994).

[81] Agent, Bareilly, to Ch. Sec. to GOB, 3 May 1811 in BC F/4/369/9221.

[82] *See* deed of transfer by one self-acknowledged 'poor woman' in letter of Commr. Nagpur to Sec. to GOB, 1 October 1841, BCrJC, 12 October 1841, no. 47; for transformation of mortgages into slave-holding in Tripura, *see* Report on Hill Tipperah for 1876 in Coxhead to Sec. GOB, 26 June 1876, BPP, September 1876, nos 30–31.

mortgage and pawnage with dishonour and discredit.[83] Malcolm had noticed this during a period of scarcity in Malwa, where 'Rajpoots, and men of other tribes, particularly Sondees, (were) selling the children whom they have by their slaves, who are considered born in a state of bondage'.[84] A poor weaver in eighteenth-century Dhaka too bequeathed a slave to his widow to transfer in times of need.[85]

Partly because these transfers occurred in non-market situations, and partly because the people transferred could be either non-slave kin or kin-termed slave, much confusion has resulted. For instance, the 'gift' of such metaphoric 'daughters' is an obvious object of reappraisal. European travellers noted that the gifting of young slaves was a well-established feature of political and economic life.[86] Where a political relationship is mediated through the gift of people-as-wealth, such a transfer could take place in various contexts. The one most familiar to us is that of the vanquished in war, but much more frequently such 'gifts' were given in order to extract some concession from the donee. W. Bolts' comment on the rajas and zamindars 'offer(ing) one of his daughters for the seraglio of an English collector, upon the adjustment of his Bundobust' becomes coherent if we keep dual meanings of kinship terms in mind.[87] Similarly, an account of the Rajas of Benares said that Mansaram obtained a *sanad* for his *zamindari* from Nawab Safdar Jang 'through Muhammad Quli Khan whom he had made his patron by making him the present of a slave-girl'.[88]

An indigenous narrative of the funeral rites of the Raja of Dhar in the early summer of 1857 noted that among gifts to eminent Brahmins

[83] For a study of Islamic theological debates on such sales, *see* M. Athar Ali, 'Elements of Social Justice in Medieval Thought', Indian History Congress, 57th Session 1996 (Delhi, 1997)

[84] John Malcolm, *A Memoir of Central India, Including Malwa and Adjoining Provinces* (London 1823–4, reprint Delhi 1970), II, p. 200.

[85] Cited in Sushil Chaudhury *From Prosperity to Decline: Eighteenth Century Bengal* (Delhi, 1995), p. 136, fn. 119.

[86] Niccolao Manucci, *Storia do Mogor,* trans. W. Irvine, (Reprint Calcutta, 1966), II, pp. 38–9; Francois Bernier, *Travels in the Mogul Empire,* trans. A. Constable, revised V. Smith (Milford, 1916), p. 135.

[87] William Bolts, *Considerations on India Affairs; Particularly Respecting the Present State of Bengal and its Dependencies,* (London, 1772), p. 162.

[88] *CPC,* V, no. 1407, p. 306.

were 'four horses, three camels, ten bullocks, nine buffaloes, thirteen dasi'.[89] Similar gifting appears to have occurred in Dinajpur in the late eighteenth century, except that the narrator appears to have been misled by indigenous terms. Describing the gifts to Brahmins on the occasion of the funeral rites of the Hindus, Walter Hamilton said that the 'wives' of the Raja had been given to the Brahmins.[90] Another report of a projected ritual oblation that went awry was that of a Vaishnavi desirous of offering her twelve-year-old *kanya* (daughter) for distribution in the *dāna-sāmagri* at the *shrāddha* (funeral) of Babu Ramdulal De in Calcutta. Finding herself too late for the event, she then sold the girl to Raja Krishnachandra Ray for a hundred and fifty rupees.[91] Apart from this ritual gifting of slaves, there were testamentary transactions upon the death of a holder, by which slaves were passed from one holder to another.[92]

Unfortunately, both the pandits and the officials of the English East India Company have been responsible for the representation of 'gifts' of metaphoric daughters as the highest form of 'marriage' where females were subjects of the transfer. In the initial chapters that went into the making of N.B. Halhed's *Code of Gentoo Laws*, amongst the five forms of 'marriage', is the Deeyb, (*daiva*) described as the 'Duchneh (dakshina)' given to a Brahmin 'procured to pray' to the household deities. In return, householders 'adorn their daughter with fine ornaments and handsome Cloaths, and give her' to the Brahmin 'in lieu of the present'.[93] The gift of the daughter valorised in current

[89] Vishnubhat Godse, *Majha Pravasa*, (D.V. Potdar ed., reprint Pune, 1974), pp. 28–9. For similar purchases and gifts of slave-women as funerary gifts in the late-eighteenth century, *see SSRPD*, pt 8, III, nos 1099 and 1104, pp. 250, 252.

[90] Walter Hamilton, *A Geographical, Statistical, and Historical Description of Hindostan and the Adjacent Countries* (London, 1820), p. 102.

[91] Benoy Ghosh, *Kolkata Shaharer Itibritta* (Calcutta, 1975) p. 542.

[92] *See* deed of transfer of the slaves of the deceased Chiranjiba Barakaistha to Rajguru Sarbananda Goswami by the Maharani of Kuch Bihar, 1783, in Amanatullah Ahmed, 'Slavery in North East India', *Indian Historical Records Commission* (henceforth *IHRC*), pt. II, no. 23, 1946. This was a predominant method of transferring slaves among the households of British officials in the late eighteenth and early nineteenth century.

[93] President and Council in Bengal to Court of Directors, 24 March 1774, in *Proceedings of the Governor and Council at Fort William Respecting the Administration of Justices amongst the Natives in Bengal* (London, 1774), p. 91.

social anthropology is not what is being defined here: the priest is in the position of a servant and moreover, the nature of her daughterhood is problematized by the fact that she is given 'as payment for services rendered'.

The anthropological discussion, following from the textual one, has concentrated on the 'gift' aspect of the transaction, rather than on the kind of 'daughterhood' or 'sonship' involved. Since prescriptive texts are silent about the various types of daughters, in contrast to their eloquence regarding various types of sons (*dattakputra, palakputra,* and *poushyaputra* to name only a few), so too are the ethnographers. If historians consider the possibility of metaphoric daughterhood then descriptions of individual and aggregate transfers of such 'daughters' gain a new significance. The biographer of an eminent arriviste family of the nineteenth century noted that a brother-in-law's mother Kshudumani had been sold 'to a well-to-do childless couple for a handful of Kshud or grain'; the latter then gave her in marriage into a family of *lathials* (armed servants).[94] Speaking of the ways in which 'brides' could be had without the payment of a heavy *pana* to the bride's father, another nineteenth century account notes,

> People take a boatload of Bangshaja girls. They may not be Brahmins at all but they are passed off as such and Bangshaja bridegrooms buy these girls from out of the boatload... poor people buy these girls. Instances have been known when Mahomedan girls sold out of the boatload have been passed off as Bangshaja girls. One such girl after her marriage ordered the servant to bring a *cherag,* which is a Persian word for lamp.... On cross-examination she was discovered to be a Mahomedan but *she was not driven away* by her husband and she remained in the family, as such a procedure would have entailed a loss of caste, and the matter was hushed up.[95]

Nineteenth century sources are replete with references to such institutions.[96] When an informant from Sylhet was asked whether

[94] Somendra Chandra Nandi, *History of the Cossimbazar Raj in the Nineteenth Century,* (Calcutta, 1986), I, p. 268.

[95] Basanta Coomar Bose, *Hindu Customs in Bengal* (Calcutta, 1929), pp. 33–4. Emphasis added.

[96] For references to 'poor Rajputs' marrying the kidnapped 'daughter of a Passi or a Mussalman', *see Annual Report on the Administration of the Province of Oudh, 1872–73,* pp. 39–40. For discussions of the economic rationality of acquiring slave-wives

such transportation and sale constituted a domestic slave-trade, she distinguished the isolation of the 'slave-wife' from other wives: whereas the latter continued to have claims upon kin for redress of wrongs and retained visiting rights, and their children had both sets (maternal and paternal) of kin, the former did not. As this informant put it, 'these girls never again in their lives saw their *atmiya swajan*' (relations).[97] Similarly, a Bengali account from the nineteenth century described the way landed and heirless men could easily procure sons from unrelated poor families, confirming this transfer through the ritual of the *poita* (sacred thread, supposed to be worn only by the twice-born castes), rearing such *bhikkāputra* as they reared calves (*bachhur poshanir moto cheleposhani*).[98] In other words, there is much in the Indian evidence to recommend the deployment of the 'slavery-to-kinship continuum' proposed by S. Miers and I. Kopytoff.[99]

These authors insist that slavery, as a *process*, rather than a fixed status of either 'unfreedom' or 'property', has to be discussed along with other institutions of 'incorporation' like marriage and adoption, without denying the exploitative dimensions of either. The Miers-Kopytoff hypothesis stresses different rates and aspects of inter-generational and intra-generational incorporation,—the affective, the jural and the economic. As a newly acquired human being, the alien recruit may only be incorporated in affect, while second or subsequent

---

in preference to same-caste brides, *see* Gurucharan Mahalanobis 'Atmakatha' in Nareshchandra and Manu Jana (eds) *Atmakatha* (Calcutta, 1982), II, p. 13, Panchanana Ghoshal, *Aparadha Vijnana* (Calcutta, 1954), III, p. 244, and Memorandum by Poll. Sec. GOBy, 23 December 1835 in NAI, FPP, 19 September 1836, no. 11.

[97] Smt. Amita Guha, personal communication of 23 September 1994, and 26 December 1996.

[98] Nistarini Devi *Sekele Katha* in Jana (eds) *Atmakatha,* II, p. 6.

[99] Susanne Miers and Igor Kopytoff, 'African "Slavery" as an Institution of Marginality' in Miers and Kopytoff (eds) *Slavery in Africa: Historical and Anthropological Perspectives* (Madison, 1977), pp. 3–81. For critical evaluations of the hypothesis, *see* James L. Watson (ed.) *Asian and African Systems of Slavery* (Oxford, 1980), Introduction; C. Meillasoux *The Anthropology of Slavery: the Womb of Iron and Gold,* trans. Alide Dasnois (London, 1991); and Jonathan Glassman, 'No Words of Their Own', *Slavery and Abolition,* 16, 1, 1995, pp. 131–45 and 'The Bondsman's New Clothes: The Contradictory Consciousness of Slave Resistance on the Swahili Coast', *International Journal of African History,* 32, 1991, pp. 277–312.

generations may move towards a more substantialized and complex incorporation within a kin-group and community. This stress on generational change also leaves open the possibility of the persistent recruitment of slaves, of the cyclicity of enslavement and incorporation, and above all of the slave of the first generation as a slave holder her/himself over time. Those who criticize the Miers-Kopytoff hypothesis for being assimilationist when slavery entailed permanent alienation and disabilities generally ignore the fact that within *each* region and social formation a multiplicity of kinship types occur. The historical evidence around Indian kinship is full of the multiple and layered meanings of identical kinship terms, and kin-forming rituals. Where slave women were 'married' plural forms of marriage existed. Certain forms by which slaves were contracted by free masters had inferior social consequences from other higher forms reserved for isogamous marriages between non-slaves. An instance of this from the Indian evidence is that of *nikah.* In classical Islamic law, this denoted marriage between two free persons. Yet in practice, in both nineteenth-century Bengal and Western India, 'nikah' seems to have denoted inferior marriages, at least one of the parties being a slave. Thus a Mussalman goldsmith who had puchased an Ethiopian female at Muscat and brought her to Kutch claimed that he had 'contracted Nika (a left handed marriage) as is the practice'.[100] Similarly, where young males were purchased for strategies of heirship, plural forms of agnation-filiation existed, with clear ritual and other criteria. This hypothesis facilitates the investigation of the plurality of kinship forms in historical time, without endowing to them any prior and fixed status, affect or benevolence. It enables us to ask difficult, perhaps inconvenient, questions like (a) Was the household a broad porous unit, made up of a variety of kin some of whom were part of the real or realizable wealth of the household? (b) Was the value and affect attributed to kinship present if there had been little or no investment in a kin-member (a newly-born infant, or a newly acquired bride)? (c) Was kinship then, a peculiarly fragile condition, without any co-relation between affect and substance, where every elder kin-member could alienate weaker and younger kin, male and female alike? In order to answer some of

[100] Translation of deposition of Rehun Toolah taken before the Political Agent, in J. Ogilvy to A. Malet, 7 June 1849, BC F/4/2343/123030.

these questions we have to turn to another missing link in Indian historiography of slavery—the ruling households.

THE HOUSEHOLD OF THE PALACE

The simultaneous elaboration of the meanings of reproductive politics and kingship in the context of slavery has not proceeded very far in the Indian historiography. The responsibility for this lies with colonial officialdom, to whom slavery as a state of agrarian and menial labour was a condition specific to men. As 'domestic slavery', especially of so-called elite households, could not be either productive or onerous, the haram too was rendered insignificant. Therefore, the reproductive aspects of medieval and precolonial polities have been largely ignored by medievalist and modernist historians alike. Part of the explanation may lie in the fact that hitherto the moral load borne by slave-holding has met the history of Indian imperial regimes between the fourteenth and eighteenth centuries head-on. The collision has been unfortunate: categorization of certain ruling houses in North and East India as 'Islamic' and associated with 'immoral' practices has served the retrograde political agendas of some[101] and worried others into silence. In the process, what appears to have been a common denominator in the constitutions of the eighteenth-and nineteenth-century ruling households in Mysore, Awadh, the Maratha, Sikh and Rajput states has been relatively obscured.[102] Just how universal slave-use was can be glimpsed in the defence offered during the mid-nineteenth century, when particular Hindu ruling houses refused to ban the sale of children on grounds such as the following:

[101] A notable instance is K.S. Lal's *Muslim Slave System in Medieval India* (New Delhi, 1994), p. 8, where an Islamic system of slavery is described as 'revolting to the Hindu psyche because it was alien to Hindu Dharma and ideologically abhorrent to it.' Tagging this sentence on to a paragraph which explicitly records the transfer of slave girls 'to the English gentlemen' at Multan in 1820s is suspect on both academic and political grounds.

[102] For the purchase of seventy-five female slaves from Delhi by the agent of Haider Ali, *see Maheshwardarbarchi Batmipatren,* no. 69, 28 July 1782, in *Itihas Sangraha,* I, 12, 1909, p. 156; Michael H. Fisher, *A Clash of Cultures: Awadh, the British and the Mughals* (Delhi, 1987) and Richard B. Barnett, *North India Between Empires: Awadh, the Mughals and the British, 1720–1801* (Delhi, 1987).

in the Zenanas of the Durbar and other Sirdars it is not customary to employ men Servants, and married women... cannot always be in attendance which is of itself a source of inconvenience—on the other hand there are Gosaees &ca who never have offspring and according to ancient custom they purchase a Boy on whom they bestow all their property...[103]

The task of reopening the debate has been considerably advanced by Sunil Kumar.[104] Concentrating on the ways in which specific male slaves were given local administrative initiative, through the *iqta* (transferrable revenue assignment) system, Kumar has extended the meanings of both 'freedom' and 'class' in such conditions. Since slaves of the sultan were ennobled, Kumar argued that freedom lay in having options, and nobility in being rooted within local kin-and-lineage groups within the empire. Since slave-status was not equated with 'landlessness' *per se*, it must have been the inability to convert assignments of revenue-paying lands into 'ancestral' lands which would distinguish the slave from the non-slave noble. It would also imply that the conversion of holdings into 'ancestral' land was only possible where the assignment holder, in turn, had a network of kin and corporate groups centred around him. In sum, Kumar's argument would lead us to conclude that the *creation* of these networks, through a range of strategies, was incumbent upon each such imperial appointee.

Kumar's argument, by extension, highlights the significance of reproductive politics in such a formation and puts the haram at the centre of a map of political culture. Admittedly, writing the haram into the given historiography requires not the erection of alternative mythologies, but a refined theoretical perspective, within which an exploration of power, articulated through hierarchies of age, wealth, social status and sexuality, can occur. It is no more fruitful to discuss Mughal political culture in terms of liberal-feminist or other philosophical paradigms than it is to ignore the reproductive bases of that polity altogether. Two disparate studies of the Mughal haram,

[103] Translated Kharita from H.H. (Gaekwar), to Lord V. Falkland, Governor of Bombay, 19 May 1850, BC F/4/2401/129609.

[104] Sunil Kumar, 'When Slaves Were Nobles: the Shamsi *Bandagan* in the Early Delhi Sultanate', *Studies in History*, 10, 1, n.s. 1994, pp. 23–52.

Findly's *Nur Jahan* and Lal's *The Mughal Harem*[105] illustrate the pitfalls in the way of such recovery. While the former details the organization of life in the haram/*zanana* complete with laws of seclusion, structures of seniority and rank, and the diversity of functions, the author's inability to relate this to reproductive politics and kingship in specifically non-European terms, the recurrent miscognition of the meanings of private and public, kinship and subordination, prohibit the theorization of a politics of lineage-making as fundamental to the processes of state-making. Similarly, Lal's work does not venture to theorize the values of wealth-creation, followership, and the formation of class and state. Lal's pre-occupation with a homogenous and hermetically sealed category, 'Muslim', makes him oblivious of the harams of other contemporaneous groups like the Rajputs. An antidote offered by Varsha Joshi proves the existence of slave-based haram systems within the ideological and ritual confines of 'Hindu' Rajput polities.[106] Albeit prodigious in its revisionist scope, Joshi's study also has its limits. Though she finds that sons born of slave-mothers did not inherit the throne, she says nothing of the 'lineage-based' state that needs slave-mothers in the first place. Nor does she scrutinize the implications of slavery for the tensions and conflicts within the nexus of kinship and kingship, and the complex role of the British colonial regime in the persistence or modification of these structures. Definitional problems remain unexamined precisely because of the 'closed' nature of lineage and caste that all these historians begin with. Equally, they translate the terms of kinship and affinity, the core ideologies of legitimation by slave-holders, in very literal ways. Thus assumptions regarding labour-forms and meanings of kinship and affinity prevent the recognition of 'marriage' or 'daughterhood' as operating within a corpus of linguistic strategies, whereby *all* the women within the princely household become forbidden to others,

[105] Elison Banks Findly, *Nur Jahan : Empress of Mughal India,* (Oxford, 1993) and K.S. Lal, *The Mughal Harem* (New Delhi, 1988); also *see* Barbara D. Metcalfe's review of Findly in 'Narrating Lives: A Mughal Empress, A French Nabob, A Nationalist Muslim Intellectual', *JAS*, 54, 2, 1995, pp. 474–9.

[106] Varsha Joshi, *Polygamy and Purdah: Women and Society among Rajputs* (Jaipur, 1995); also Marzia Balzani and V. Joshi, 'The Death of a Concubine's Daughter', *South Asia Research,* 14, 2, 1994, pp. 136–62.

yet are distinguished from non-princely daughters and wives in diverse ways.

Why the Nazims of Bengal?

The revision of false universals—the dichotomy between public and private, between equality and inequality, or wealth-holding and landlessness—need to be the fundamental premise of any reappraisal of precolonial polities in India. It is with this revisionist intent that I approach the history of a specific household: that of the Nazims of Murshidabad. The choice of the Nizamat and Bengal is dictated by its obvious pre-eminence both as an outpost of the Mughal empire and its simultaneous role as the fulcrum of the East India Company. Both these aspects are crucial to understanding the transition between a pre-colonial political economy of slavery to a colonial economy based on slave-labour. The first aspect, the Nizamat as a Mughal formation, was important because even in the early nineteenth century, a Nazim at Murshidabad offered one hundred and one gold mohurs to a newly proclaimed Mughal emperor, proclaimed the accession by beat of the *naubat* and issued new seals with the name of the Emperor. An extremely vigilant Company authorized such actions as 'consistent with established usage... observed by the Court of Murshidabad'.[107] The records of the eminent households of the region all argue for the persistence of ideals and structures that were identifiable as that of a Mughal Islamicate elite.[108] The shared emphasis on conspicuous consumption by which status was displayed was common. In 1816, the Company's servant Samuel Droz reported that his elephant passing the Nandi residence was accosted by the scion and a particularly abusive 'Caffree boy' in the company of the Kuar.[109] The skilled performers—the dancing and singing Ashūrans and Misrīs—were equally common:

[107] Persian Sec. to AGG, 28 January 1807, upon the accession of Akbar II, *Murshidabad Nizamut: Letters Received,* 1807–1855, (eds J. Datta Gupta and S.K. Bose, Calcutta, 1969) henceforth *MNLR,* II, p. 1.

[108] I use the term to refer to hegemonic social and cultural standards, rather than a religious unity. For a recent discussion of the Islamicate, *see* Philip B. Wagoner '"Sultan among Hindu Kings": Dress, Titles and the Islamicization of Hindu Culture at Vijaynagara', *JAS,* 55, 4, 1996, pp. 851–80.

[109] S.C. Nandi *History of the Cossimbazar Raj,* pp. 46–7.

hence the Durga Pujas in the houses of the arriviste zamindars like the Mullicks and the Debs in early-nineteenth-century Calcutta were archetypal occasions for the display of this shared consumption.[110] In many instances such women would be considered—and consider themselves—as part of the household. Hence alongside the petitions of the sister and four wives of Raja Tejchand of Burdwan claiming certain land and immoveables is that of 'one of the women of the Haram' claiming a garden and *pakka* (built of bricks) house in Burdwan.[111] Many of the conflicts around slave-concubinage and slave-birth and succession would thus be common to these households in the nineteenth century. Hence, if we could comprehend some of the ways in which slaves functioned in at least one of these households, it would provide us a paradigm for other similar households in the region. A study of the Nizamat might bring home both the particular history of one such slave-based formation and point out its shared fate with other such households.

Furthermore, though it is hard to appreciate this fact in the late twentieth century, the Nizamat loomed large in the historical imagination of many in the nineteenth century. As a report of the late nineteenth century on the Nizamat stipendiaries put it,

> a considerable degree of political prestige seems still to surround them; not only in and around Moorshedabad itself, but in whatever parts of Bengal they enter in pursuit of sport; and the cordiality with which the shopkeepers of Moorshedabad, and even the peasantry of the surrounding country, do reverence to the members of the Nizamat family in their daily rides and drives, is an indication perhaps worthy of being noted. The Agent to the Governor-General may be riding by the side of the young Nawab; but not till the latter's eye has been caught and a low salaam been made to him, is so much as a glance turned in the direction of the representative of the Government.[112]

Histories of local families of affluence written in the late nineteenth century continued to register their proximity to power by virtue of

[110] H.Sandeman (ed.) *Selections from the Calcutta Gazettes, 1806–1815,* (Calcutta 1868), IV, p. 370.

[111] Letter of Khaunim Bebee, in A. Mitra (ed.) *West Bengal District Records: Burdwan, Letters Received, 1788–1802* (Calcutta, 1955), p. 307.

[112] 'Report on Some of the Political and Historical Features of Moorshedabad in 1876', AGG Tweedie to Sec. GOB, 31 August 1876, BPP, September 1876, no. 9.

affinal relationships with the Nawab Nazims of Murshidabad. For instance, a history of the Nahars of Azimganj (Oswal Jains) narrated that one of the scions of this dynasty had a father-in-law who had 'stolen a *nabobputri* from Murshidabad', married her according to the *sharia* and thus been pardoned the theft. The family thereafter enjoyed a special relationship as the *douhitra* (daughter's heirs) of the Nawabs.[113] Other local histories also treated affinity with the Nawabs as a yardstick of prestige or lack of it. Thus a local history of the Rajas of Nadia recounts the giving of a Kulin Brahmin's daughter to the Nawabs of Murshidabad as a preferred choice over the gifting of the girl to a lower-caste (*kishorkuni*) Krishnachandra.[114] On the other hand, numerous stories and folk-songs celebrate the resistance of provincial Hindu zamindars to yielding up their daughters to the Nawabs.[115] However represented, affinity with Murshidabad looms large in the local histories. There is then, an even greater need to historicize the ways in which the household was constituted within the Nizamat.

The second aspect of the Nizamat, as a fulcrum of colonial intervention, is important for the articulation of values and structures that the East India Company implanted in the course of the nineteenth century. Most critically, studying the Nizamat suggests the limits of colonial agendas regarding law, both indigenous and Company Regulation. Since slaves were important to both precolonial and later colonial formations, this study attempts an understanding of the refashioning of one by the other.

[113] Avinashchandra Das, *Nahar Vamsa Vrittanta* (Calcutta, 1895), pp. 24–7.

[114] Kumud Nath Mullik, *Nadiya Kahini* (Calcutta, first 1317 B.S./1910, consulted 3rd. edition 1986), p. 234.

[115] For an example, *see* the ballad of Rupavati in Dinesh Chandra Sen, *Ballads of Bengal* (first published 1923, reprint Delhi 1988), I, pp. 187–203.

2

Kinship and Kinlessness in the Nizamat of Murshidabad

On 1 November 1880, by an indenture signed between the Nawab Nazim of Bengal, Bihar and Orissa and the Secretary of State in England, the Nazim Farīdūn Jah agreed to withdraw from affairs of government, in return for a provision for two of his sons and two daughters,[1] and was formally transformed into a 'mere creature of sufferance' at Murshidabad.[2] Historians have generally taken 1770 as the cut-off mark[3] to suggest that since political and economic power passed to the Company, the household of the Nizamat ceased to be politically or socially significant. This is an oversight for if the Nazims had been effectively marginalized by 1770 or by 1793, then the simple act of a Nazim conferring a title on a descendant of the Jagat Seths in 1814 would not have merited the displeasure evinced by the government in Bengal. His power was 'utterly incompatible', it was said, with the aims of the Government which wished to observe 'all the forms of external respect... but not... to allow him to exercise any acts calculated to effect its own real supremacy'.[4] Nor would another English official report that the Nazim

[1] For copy of the agreement *see* L/P&S/18/D 40.

[2] Note by Dalhousie, Gov. of Bengal, 22 April 1854, NAI, FPP., 26 May 1854, no. 82, folio 20.

[3] D.N. Banerjee, 'The Accession of Nazm-ud-Dowla to the Throne of Bengal and the Position of the East India Company', *BPP*, 60, 1941, pp. 19–34; Niharkana Majumdar, 'Bengal Politics in 1765', *BPP*, 75, 1956, pp. 98–117; Nanigopal Chaudhuri, *Cartier, Governor of Bengal, 1769–1772* (Calcutta, 1970).

[4] Poll. Letter from Bengal, 15 June 1814, BC F/4/426/10, 409.

had visited him one evening in 1838, offering him three lakhs of rupees for a crown, on the grounds 'that we had made a King of Oude, and could make one of Bengal'.[5] The erosion of power and wealth was gradual and involved the deployment of many different policies on the part of the Company. Many of these policies and strategies regarding the affects of kinship and the significance of law transformed a slave-based polity into a decaying 'family'.

The major instrument of these transformations was economic. By the Treaty of Allahabad, the Company was to pay fifty three lakhs of rupees annually to the Nazim. The Company while committed to upholding the forms and rituals of the polity, was financially committed to stockholders in Britain. However, mismanagement, and erosion of profits through famine and other related factors led to the drastic reduction of the amount paid to the Nazim by 1771. The stipend of the minor Nazim, Mubarak-ud-daula, was reduced to sixteen lakhs in 1770 and never restored to the amount promised in the treaty of 1765. With this began the policy of retrenchment, which intensified as the wars at the turn of the eighteenth century landed the Company in financial crises. Thus between 1773 and 1818, there were at least five committees appointed by the Government in Bengal to scrutinize the minutiae of expenditures of the Nizamat.[6] The monitoring of Nizamat expenditures was directly related to the Company's efforts to create a Deposit fund, entirely in the management of its officials, and made up of 'lapsed' stipends and allowances. The Nizamat Deposit Fund, created in 1818, provided the Company with the single most important lever with which to refashion the Nizamat. From these funds came the salaries of various English military officers and civil servants of the Company, resources for building new palaces,[7] a Nizamat College, experimental horticultural gardens and a new Imambara. Thus while these Deposit Funds allowed the Company to

[5] AGG Caulfield to Sec. to GOB, Prinsep, 8 August 1838, *Murshidabad Nizamut, Letters Issued*, 1834–72 (ed. J. Datta Gupta and S.K. Bose, Calcutta, 1967), henceforth *MNLI*, II, p. 231.

[6] For a summary of the Government's attempts over the period 1790–1816 to establish and maintain 'a due restraint over the disbursements of the Nizamut' *see* Note by N.B. Edmonstone, 25 March 1816, BPC, 23 July 1816, no. 1.

[7] Biswanath Roy, *A Peep Through the Hazarduary Palace of Murshidabad* (Berhampore, 1981).

pay itself, it also enabled it to behave as a 'feudal overlord' redistributing largesse and loans to various members of the Nizamat itself.

Every retrenchment of expenses, and every addition to the Nizamat Deposit Fund, was guided by an ideology of kinship as derived from European aristocratic models. This chapter attempts to outline some of the principles along which the household-polity was organized and functioned. The first two sections are devoted to an examination of the relationships in the predominantly 'male' world of the Nizamat, and the potentials for conflict within this world. These conflicts and relationships, in turn, represented the playing out of the hierarchies refined and reproduced in the inner world of the Nizamat. The following sections examine the structures of the haram, and the material economy of the household upon which the Company's Agents tightened their grip through the reworking of controls in the 'outer' spheres of the polity.

When Kin were Enemies: The Nazims, Kin and Slaves

Power in Islamicate polities was imagined and arranged in terms of concentricity, and represented by spatial division more horizontal than vertical. Hence the importance of the inner as the centre of this series of concentric circles, identified with the ruler, and the space he inhabits. Etymologically derived from the Arabic root *h-r-m*, the term haram combines two sets of meanings, one, to be forbidden or unlawful, and the second, to declare sacred and inviolable. Thus a haram is, by this definition, a sanctuary or a sacred precinct, redolent of religious purity and honour, and evocative of the requisite obeisance. By implication, it is a space to which general access is forbidden or controlled and in which the presence of certain individuals or certain modes of behaviour are forbidden. Thus proximity to the ruler, and access to interiors, determined rank and power. Rewriting the spatial representation of power is fundamental to an appreciation of regimes, like those of the Mughals, where the person of the ruler creates a sacred space around the presence. In such regimes, only those who were not considered to be fully adult males could be routinely permitted in the inner worlds of the palace: eunuchs, dwarfs and mutes, and slave-women and children of both sexes. Thus eunuchs and female slaves could hold the highest offices, with corresponding salaries and incomes. It is within this political hierarchy, an inversion of dominant Western notions of the politically significant as 'outer' or public, (and male) and the politically marginal as 'inner' or private and domestic, that the meanings of seniority, gender and power can be understood.

As Eaton notes, the model of imperial authority projected from the Mughal court represented the sovereign as the centre and apex of virtue, order and dignity.[8] The symbolic translation of this political universe rested upon a minutely regulated access to the person of the ruler. Duplicated in miniaturized form in Bengal from the time of Jahangir, the governor's *jharokha* also consisted of a window and enclosed space built on a platform elevated some twelve feet above the ground where he received those admitted for audience. Also in imitation of Mughal practice, the precise position of each member relative to others would be indicated by the distance from the jharokha—those closest were understood to be high in the ruler's esteem. Contemporary observers described the fastidious observance of decorum by the 'great men' assembled to pay their respects to the Nazim.

The Europeans were placed near the Nabob's person... directly opposite to the Nabob, a fountain was playing. The Moors who entered, approached no nearer than just before the fountain, there made obeisance, and then returned to their seats.[9]

At other times, when the ruler journeyed from the palace, for instance, an equally elaborate system of avoidance and deference was expected from subjects. An uncomprehending English surgeon reported that on his way from Murshidabad, the reins of his horse were seized by the servants of the Nazim, who insisted that he

must not proceed further but dismount, and remain by the road side, till His Highness who was coming that way, had passed by... the Nawab passed me in an open palanquin or Nalkee, and being personally acquainted with His Highness, I addressed him, but he either did not or would not hear me.[10]

[8] Richard M. Eaton, *The Rise of Islam and the Bengal Frontier, 1204—1760* (London, 1993), p. 160.

[9] Orme, *Orme's Historical Fragments of the Mogul Empire, of the Morattoes, and of the English Concern in Indostan* (1782, Reprint Delhi, 1974), pp. 273–74.

[10] R. Burt to G. Swinton, 20 July 1830, *MNLR*, II, p. 213.

LEEDS UNIVERSITY LIBRARY

What the surgeon had unwittingly observed was the symbolic expression of a code which both projected and protected the seclusion and sacredness of the person of the Nazim. However, proximity in this code was not always enjoyed by kinsmen. Many of the contests over the symbols of honour and power occurred not between strangers and the Nazims but between those who in the Western European sense of kinship, were closely related to the Nazims. An illustration of this is the effort of the Nawab of Dacca, whose younger brother Shams-ud-daula was married to the daughter of the Nazim Mubarak-ud-daula I, to obtain permission to use a *nalkī*. Upon the Company presenting him with one, the reigning Nazim protested

> except that the Nobab Muzuffur Jung as Naib of the three Soubahs of Bengal, Bahar & Orissa [...] obtained the Royal sanction for using a Nalkee to this time neither the Subahdar of Patna nor the Naib of Dacca ever asserted Pretensions to this mark of Distinction.... Nussurut Jungs Position is not such as to give him a claim to such high token of Consideration.[11]

Similarly the erection and use of the naubat by the elderly Wālida Begam was interpreted both by the Nazim and the British Agent as a challenge to the authority of the former.[12] Or, as in 1824, the very fact that the head of the Mahalserai (apartments of women), Amīrunnisa Begam, had fired her musketry after the firing of the Nazim's artillery and musketry to observe the ceremony of 'Id, aroused the Nazim to accuse her of 'insulting his dignity by infringing a rule of the deepest importance'.[13] On this occasion, the chief obstacle to a reconciliation between the Nazim and the Begam became the terms of address to be used by the latter: to the 'astonishment and serious vexation' of the Agent, she refused to send her 'Bandagī' to the Nazim, considering this mode of due submission, 'so great a degradation' that she threatened to die if it was insisted upon. She preferred instead sending 'Salām', which in the words of the Agent denoted 'the mode of an address that of an equal'.[14]

[11] Translation of a letter from the Nawab, 2 January, 1809, *MNLI*, I, pp. 117–8.

[12] G.Swinton to AGG Martin, 30 September 1818, *MNLR*, II, p. 163.

[13] W. Loch to A. Stirling, Persian Sec. to Govt., 14 August, 1824, *MNLI*, I, p. 307.

[14] Ibid.

These matters were issues of life and death precisely because relationships of blood and affinal kinship could prove dangerous. The insignia of status were closely monitored because succession stayed open in the first three decades of the nineteenth century. Older patterns of accession to the throne exercised an influence on the minds of many potential contenders, like brothers and uncles. The possibility of succession devolving collaterally, that is by transmission of authority to the next-born of elders within the same generation before passing on to the elder of the next generation, remained an ever present threat to both the Company and the young Nazims it chose to nominate from time to time. As the brothers of a deceased Nazim reminded his son,

> 'we are not among the dependants or distant connections of the Nizamat. We are the sons of His late Highness... & are therefore in the first degree of relationship to this House. And every one of us, in the event of fortune so favouring him, has a right to succeed to the masnad'.[15]

This was no empty threat: three brothers had succeeded each other on the *masnad* of Murshidabad between 1765 and 1770—Najm-ud-daula, Saif-ud-daula, and Mubarak-ud-daula. Upon the death of the latter in 1793, it was true, his eldest son, Babar Áli, had succeeded him. But this lineal succession was possible only because the ruling female Regent, Manni Begam, had overruled Mubarak's own preference for a younger son, Abul Qasim Khan (better known as Mīr Mangli). However, when Babar Ali died, this matriarch recommended the 32-year-old Manglī to the throne, in preference to a son of the deceased Nazim. She cited, as precedent, the patterns of collateral succession in the Rajmahal and Purnea families and also that of 'the present Nabob of Oudh, Saadat Ali Khan'.[16] The Government of Bengal defended the principles of lineal succession and primogeniture on the grounds that it was according to the 'usages of Hindostan', the 'principles of the Mussulman Law' and the 'practice of all countries, Governments and Principalities'.[17] Of such contortions, Lord Hastings observed in 1814

[15] Enclosed in R. Rocke to P. Monckton, Pers. Sec. to Govt., 15 November 1810, BC F/4/372/9258.

[16] Letter from Manni Begam, recd. 17 May 1810, enclosed in Minto and Colebrooke to the Court of Directors, 30 May 1810, BC F/4/312/7142.

[17] N.B. Edmonstone, Chief Sec. to Govt., to Manni Begam, 26 May 1810, ibid.

We constantly oppose our construction of Mahomedan law to the right which the Moslem princes claim from usage to choose among their sons the individual to be declared heir-apparent. It is supposed that by upholding the right of primogeniture we establish an interest with the eldest son which will be beneficial to us when he comes to the throne... nothing can be more delusive... some interest of the Company is always really involved.[18]

Whatever its interests in promoting an infant to the masnad in 1821, the Company's government reverted to collateral succession again, when the Nazim Alijah was succeeded by his brother Walajah. Lineality and primogeniture were not definitively established in Murshidabad till Walajah's grandson, Farīdūn Jah, was appointed Nazim over the claims of his grand-uncles in 1838. The Company's efforts at fixing a principle for succession arose from English notions of descent, consanguinity as well as ideological and pragmatic concerns.

Succession alone did not create antagonism between Nazims and their male kin. Control of lucrative posts and offices was equally important. For instance, while Mīr Ja'far and Mubarak-ud-daula continued to strengthen their faction with sons-in-law and brothers-in-law from one family, (Mīrza Daùd was the son-in-law of Mīran, first-born son of Mīr Ja'far, Daùd's son Khalīlullah was the son-in-law of Mubarak-ud-daula I, and his Diwan, and Khalīlullah's son, Mīrza Jalil, was also brother-in-law of Mubarak-ud-daula), the same affinal kin fell foul of the succeeding Nazim, Babar Ali, in 1794. Accusing Khalīlullah Khan of trying to place another of his brothers on the masnad, the Nazim Nasir-ul-mulk charged the elderly affine of having violated the 'established customs of courtesy', of not signing the pension bills, of not giving in his accounts, and in a final offence, 'In spite of his position as a servant Khalil Allah Khan addresses him (writer) as his equal'.[19]

Similarly, Manni Begam, who sought to procure the office of *Mīrsamani* to the Nizamat for the son of Syed Ali Khan, the previous holder, whom she described as 'nearly related' to the deceased Nazim, Mubarak-ud-daula I, found it bestowed on Iman Quli Khan, who she claimed had no 'claims of service or long dependence upon this

[18] Francis Rawdon Hastings, *The Private Journal of the Marquess of Hastings* Marchioness of Bute (ed.) (London, 1858) I, pp. 48–9.

[19] *CPC*, XI, no. 115, pp. 24–5.

house'.[20] British intervention in the struggle between the Nazims and the respective heads of the Mahalsarāi was based on a significantly different ideological ground.

The Governor-General's instructions to Nasir-ul-mulk conveyed the range of aspirations that the Company entertained for the institution of kinship. The fabric of kin-based European polities required that administrative posts and significant emoluments went to those who theoretically claimed relations of affinity or consanguinity to the rulers. The fact that various kinds of male kin and affines in all the different accounts of the Nizamat between 1773 and 1834 appeared, as *sahibzada* and *aqriba*, to have monetary claims upon the Nazims, made them appear to British officials the appropriate figures to perform 'functions' commensurate to their stipends. Yet virtually no polity in eighteenth-century India neatly fitted European aristocratic paradigms. As the Nawab of Farrukhabād put it in 1787, while refusing to make his brother the Diwan of his administration and preferring his brother-in-law, 'it is an established custom among all the chiefs of Hindustan not to invest a brother or a son with the management of their affairs nor to entrust any of the said persons with the office of the naib.'[21]

Furthermore, as the descriptions offered by the Nazim illustrate, it was the content and structures of kinship itself that were deviant. Referring to Mirza Koka, whom the Governor-General had objected to, as my 'late father's foster brother' and subsequently as 'the son of the Nabob Mobarukuddowla's Nurse', the Nazim justified the latter's proximity to the throne on the grounds that 'He has been bred up in this family from his childhood, always attended on my late father and now continues to attend on me... in consequence of his long dependence upon this family'.[22] Having failed to convince the Nazim to appoint his 'relations' to posts in the Nizamat administration, Shore wrote to Manni Begam darkly threatening 'to adopt other measures' to bring the recalcitrant Nazim around.[23] These measures were fully

[20] John Shore to J.E. Harington, 3 February 1795, *MNLR*, pt. I, 1793–1856, pp. 15–16.

[21] Diler Himmat Khan to Council, *CPC*, VII, no. 1112, p. 312.

[22] From Nasirul Mulk, May 1795 (covering letter missing), *MNLR*, I, pp. 22–23.

[23] Enclosed in Edmonstone to Harington, 19 August, 1795, *MNLR*, I, p. 25.

worked out under subsequent Governors-General, beginning with the appointment of a Committee to enquire in to the distribution of stipends in the Nizamat in 1800–1. The copy of instructions circulated among the members of that Committee[24] laid down the principle that

> the credit of the family is concerned in giving a preference to those who are connected with it by the ties of consanguinity or who are related to the predecessors of the late and former Nabobs over persons who derive their Pensions solely from favour.[25]

The Nazim's privileging of those 'bred' in the family, over and above blood-ties, was incomprehensible to British officials. Few understood the reasons for the animosity they found between the Nazim and the other members of the Nizamat. This bloodless contest among those who were deemed to have shared blood involved a range of strategies, from the withholding of pensions and allowances, withdrawing exemptions and privileges, razing homes and even physical confrontations between groups of personal bodyguards and attendants—all of which created an 'incessant succession of broils and confusion'.[26] Most typically, it centred around the refusal of deference, which was related to the absence of personal attendance at the Court. It was said of the 'uncles' of one Nazim that they 'were

[24] They were Thomas Pattle, Chief Judge, and President of this Committee, Edward Colebrooke, Richard Rocke, also judges of the Court of Appeal and Circuit for the division of Murshidabad, and Thomas Hayes, Collector at Murshidabad. Simultaneously, T. Pattle was appointed the Superintendent of Nizamat Affairs. His brief was to be concerned with 'such portion of the Nizamat stipends as is now disbursed or to supervise the payment of the stipends and to see that they are regularly and properly disbursed to the several persons to whom they are payable', and 'to interfere for the accommodation of disputes and dissentions [sic] among the Dependants and Servants of the Nizamut' and between them and the Nazim and the Manni Begam, and of 'differences which may arise among the members of the Family'. Extract BPC, 22 October, 1801, BC F/4/101/2376.

[25] Edmonstone to Pattle, Colebrooke, Rocke, Hayes, 24 December 1801, ibid.

[26] J. Ahmutty to A.Stirling, Offg. Persian Secy. to Govt., 17 September 1822, BC F/4/851/22643. For individual confrontations between sons of Mubarak-ud-daula and the Nazim Walajah, *see* BPC, 11 April 1822, nos 131–2 and 24 April, no. 21; 7 June 1822, nos 58–60; 27 February 1824, nos 45–7. For confrontations between Nazim Mubarak Ali Humayun Jah and other male kin *see* BC F/4/1475/57976 and F/4/1522/60090A.

excluded from the presence, having the misfortune of labouring under His Highness's displeasure'.[27] Yet another Nazim, determined to defend his dignity, requested the Agent to direct his relatives 'to pay the respect due to him'. Upon this, the latter reiterated the ambition entertained by Shore, 'of his having his relatives and the members of the Nizamat family occasionally about him, which would secure their paying him voluntarily that respect which they might not yield in acquiescence to the Agent's order'. But the Nazim would have none of that, declaring that 'he did not care about them, that they were all his enemies'.[28]

Such tensions between blood-related kin created the need to have closest to the Nazim those who depended on him alone, and not those who could base their claims to the masnad by virtue of the depth of their descent from previous Nazims, or even from other ruling princes. The best qualified for such proximity were therefore the slaves of the house, from whom deference could be extracted without any loss of honour by them. Slaves alone could provide allegiance to the ruler that his natal kin did not owe him. It was because they 'owed' their lives to the ruler that they could be employed to guard him from the assaults of the enemies within. In Awadh too, the practice of deploying chelas (slaves) raised in their masters' households was premised upon their kinlessness: having no *asabiya*, or group feeling of their own, they were expected to adopt the asabiya of their master.[29]

Few Company officials realized the importance of the sanctity of the person of the ruler and how this determined the function and nature of slaves. This incomprehension was evident in the efforts of Cartier and the Council at Calcutta to diminish the *sawār* (bodyguard) of the Nazim: Clive had referred to them as the 'military rabble' but the Nazim had described them as 'old confidential servants... fixed upon the Doors of the Mahul'.[30] The composition of the sawār of the Nazims has never been studied in detail, but from the contemptuous

[27] Welland to T. Brooke, 21 February 1812, *MNLR*, II, p. 74.

[28] AGG Caulfield to H.T. Prinsep, Sec. to Govt., Poll. Dept., 8 August 1838, *MNLI*, II, p. 230.

[29] Michael Fisher, *A Clash of Cultures*, pp. 53,109–10.

[30] Cited in Abdul Majed Khan *The Transition in Bengal*, 1756–1775: *A Study of Saiyid Muhammed Reza Khan* (Cambridge, 1969), p. 286.

terms used by the officials of the Company while referring to them, it seems probable that a certain proportion of the sawār were also slaves. One reason for inferring this is the details of the dispute of 1795, when John Shore vaguely referred to the 'persons of low rank and characters' upon whom the Nazim Mubarak-ud-daula II (Babar Ali, Nasir-ul-mulk) had 'bestowed Khillauts, Titles and allowances' and thereby exceeded in his monthly charges the amount of his income.[31] Some of the men referred to were slaves, and listed under the heads of 'Shāgird Pesha' and 'Mulāzimān i-Imtiyāzī'. For instance, there was Muhammad Hellall [Hilahil], described by the Nazim Nasir-ul-mulk as an 'ancient Chela (or Domestic) and faithful servant of my Grand Father... all the receipts for salaries &ca. are authenticated by his seal, all the Sunnuds, Jewels and the Privy Seal remain under his charge'.[32] Unfortunately, the clarity of this description of the holder of the post of Darogha-i-Khazana is not paralleled for all the other posts and offices that male slaves held in the Nizamat. So the accounts of 1773 and 1802 list numbers of slaves paid from the Nizamat *behla* accounts—the latter listing ninety-three Hindustani chelas, thirty-three 'Coffree' slaves and three other kinds of slaves—though the specific offices held and performed by them are rarely mentioned alongside. Nevertheless, the physical proximity of male slaves to the Nazim, in the most intimate situations was evident in descriptions of one official of the final moments in a Nazim's life:

> His Highness's last directions were that those Khwasses who had always attended upon him should wash and purify his corpse, and after clothing him, with the binding sheet of Karbala, should inter him.... He desired the Khwasses to fetch clean clothes... to remove out of sight the close stool.[33]

BETWEEN MALE AND FEMALE: ANDROGYNOUS ANTI-KIN

Many offices were however explicitly associated with slave-eunuchs, or *Khwājāsarai*. Though associated in lay minds with Islamic polities alone, the fact was that eunuchism, like slavery in general, had existed in

[31] Shore to Harington, 3 February 1795, *MNLR*, I, pp. 15–16.

[32] Translation of remarks of the Nawab Nazim on the statements of proposed retrenchments submitted by the Nizamut Committee, n.d., Doc. F, in BC F/4/250/5602.

[33] Ahmutty to Sterling, 17 September 1822, BC F/4/851/22643.

classical Rome and Greece, as well as in ancient China,[34] long before the rise of Islam.[35] Nor does it appear that Islam 'endowed the employment of castrated men with its own special character'.[36] This is important to emphasize particularly in the context of Indian history in the eighteenth and nineteenth centuries. If the presence of eunuchs was an indicator of Islam, then what explanation can we offer for the many Hindu eunuchs in the Rajput polities in western and northern India?[37] The persistent association of different kinds of eunuchs with specific offices in the administration of various Indian polities at the same time was no coincidence. The divinization of the ruler and the ritualization of a court could only be mediated by those who were neither god nor human, neither fully male nor female, neither adult nor child. The visibility of their (eunuchs) power was explained in terms of the psychological impairment that followed upon castration. As an English judge, referring to the charges of cruelty against five eunuchs of the Nazim's Camp, put it:

> The *publicity* of the acts... may be attributed to the *presumption*... that as members of His Highness's household they had *within* his camp the power to apprehend and punish for theft, and to the *gratification* which some of them, peculiarly circumstanced as they are, might naturally feel in openly displaying that power....[38]

[34] *See* Mary M. Anderson, *Hidden Power: the Palace Eunuchs of Imperial China* (Buffalo, 1990): Taisuke Mitamura, *Chinese Eunuchs: The Structure of Intimate Politics* trans. C.A. Pomeroy (Vermont, 1970) and C. M. Wilbur, *Slavery in China during the Former Han Dynasty 206 B.C.– A.D.25* (Chicago, 1943).

[35] Keith Hopkins, *Conquerors and Slaves* (Cambridge, 1978), pp. 172–96.

[36] Ehud R. Toledano, 'The Imperial Eunuchs of Istanbul: From Africa to the Heart of Islam', *Middle Eastern Studies*, 20, 1984, pp. 379–90. The phrase seems to refer to the widely held view of the eunuchs acting as guards of the haram, suited by their castration, for the task of protecting the celibacy of the inmates. This is an over-simplification.

[37] *See* reference to the 24 *nazir* eunuchs in the Jat ruling house of Bharatpur in Jadunath Sarkar's translation of the Persian history authored by Franz Gottlieb Kuhn, *BPP*, 75, 1956, pp. 71–86, and to the *panchkanya* (castrated slaves) in the ruling houses of Rajasthan in Varsha Joshi, *Polygamy and Purdah*, pp. 172–74. For older pattern of eunuch-tributes, *see* Gavin Hambly 'A Note on the Trade in Eunuchs in Mughal Bengal', *Journal of the American Oriental Society*, 94, 1, (1974), pp. 125–30.

[38] J.D. Money, Sessions Judge, Murshidabad, to Registrar of the Sadr Nizamut Adalat at Calcutta, 22 September 1853, in *Indian Records*, p. 187. Emphasis in original.

Western observers did not appreciate the importance of androgyny and ambivalence as political attributes, making these slaves powerful agents in enforcing the particular will of a Nazim or of a head of a *de'orhi.* Hence as a visitor to Murshidabad the Earl of Moira was

> surprised at the insolent authority with which the head eunuch (who was on our side of the curtain) spoke to the poor woman [Bahu Begam]. She probably had not understood something which Mr. Brooke [the Agent] said to her on my part, on which the eunuch in a high voice and tone of reprehension asked her why she did not answer. Remarking this to the gentlemen next to me,.... I was told these fellows were allowed to treat the women with great harshness.[39]

As with the governor-general, many first-time observers failed to distinguish between the particular de'orhi that an eunuch represented. Nor did they explain that the eunuchs were the embodiment of commercial and political 'agency' for law-keeping and therefore punishment, for wealth generation and its management and for gathering information on the basis of which decisions were taken. The persistent association of eunuchs with cruelty, and violence,[40] must be understood in the context of the special ability of these androgynous beings to breach the various boundaries separating male from female worlds, the world of the living and of the dead,[41] of the Nazim, the de'orhi heads, and the populace in general.

Describing the scene at the annual festival of Berah one newspaper reported the presence of 'a few tall white robed eunuchs, whose presence, as they wandered from group to group called down a low and general obeisance from courtiers, whose polished manners and graceful salaams told how well versed they were in all the etiquette of the durbar'.[42] Or as another AGG translated in 1837, it was this deference shown to eunuchs that prevented clashes between the British-controlled police and others in Murshidabad from becoming riots, and helped to quell the 'disorders and quarrels... among the Female

[39] F. R. Hastings, *Private Journal* I, pp. 84–5.

[40] For another Agent's suspicion of the eunuch Nazir of Bahu Begam, Jawaharali Khan, for murder, see Magt. Berhampore to AGG, 23 July 1844, *MNLR,* II, p. 497.

[41] For a subtle analysis of the multiple conditions embodied by eunuchs *see* Shaun Marmon, *Eunuchs and Sacred Boundaries in Islamic Society* (New York, 1995).

[42] *Moorshedabad News,* 22 September 1838.

establishments'.[43] Clearly this deference reiterated the special power these eunuchs exercised on behalf of their particular de'orhis.

The standing of the immediate master or mistress differentiated one eunuch from another. This view is warranted by official accounts which sometimes listed eunuchs according to the hierarchy of the de'orhis. The rank of the eunuch was inferrable from the rank of the persons s/he served. The earliest reference to eunuch-officials in the Nizamat records is in the household list of 1772, where the eunuch Ambar is listed as Darogha *Mashālkhana*. Another eunuch, Ghulam Muhammad Khan is described as Darogha *Mīr Shikārān*.[44] Eunuchs in later lists also held offices like Darogha *Nawarrah* (boats). The office of the *qalamdānkhana*, through which the wealth of the house was governed, was always filled by eunuchs of the Nazims. These officers received the stipends and wages of all the junior servants from the Nizamat treasury, supervised accountants, and administered the profits arising from trade and the collections of the various revenue-farms. As one Nazim wrote of eunuch Siddi Ambar Ali Khan, Darogha of the qalamdānkhana, he had been appointed as 'tutor to both my father and uncle, had the management... of their pecuniary affairs'.[45] The treasuries of particular de'orhis were also in the custody of the eunuchs of the particular *mahalsarāi*. Thus, it was reported that the eunuch Basant Ali Khan and Zamurrud, the head eunuch of the Dulhin Begam's de'orhi, had received charge of 2200 *bhurries* of silver from the department of the *khas tehvil* (treasury) and deposited the articles in the mahalsarāi according to the orders of the Nazim.[46] The person who received the articles of gold and silver in Bahu Begam's de'orhi had also been the head eunuch of that establishment, I'tbār.

Company officials periodically protested against bestowing trust, titles and honours on eunuchs. While a Governor-General like Shore might believe that these men 'neither by station or merit are entitled

[43] Extract from AGG Melville's Despatch of 20 May 1837 in BC F/4/2432/133218. For a collateral male who desired the services of the eunuchs of the Nizamat to stop the matriarch of his household removing property from a house, *see* AGG to Dy. Sec., 16 May 1834, NAI, FPP, 22 May 1834, no. 92.

[44] List of persons under the head of Khajaserai in BMisc.C, P/154/38, 23 January 1773.

[45] Nazim to AGG, 7 Jamadusani 1224 Hijri, (1829), BC F/4/1228/40155.

[46] J. Ahmuty to A. Sterling, 17 September 1822, BC F/4/851/22643.

to the honour of your confidence', the reply of Nazim Nasir-ul-mulk illustrated the very different criteria by which such positions of honour and trust were bestowed. Arguing that the eunuch 'Seedy Amber... is an old inmate of the household', the Nazim pointed out that he 'had been entrusted by my late Father with the charge of the kullumdaun khana department. I have now out of consideration for his length of services (for there is none among the Khauja Serays of longer standing or more trustworthy) appointed him Nazir of my honoured Mother's Dewry'.[47] From this flowed the elevation of rank and perquisites, as he put it, 'to give him a greater degree of Respectability, I according to the custom of this house conferred on him the title of Khan'.[48] Titles such as 'Nawab Nazir' and Khan Sahib were conferred on the distinguished eunuchs; others were simply Miyan (address of respect). The solidity of this custom was attested to by the request of the mother of the Nazim in 1841, Ràisunnisā Begam, regarding the eunuch in charge of the business of her de'orhi, Miyān Feroz, in which she said,

> It is a long established custom in this family for the Nazir [Nazim] to bestow on the person appointed to the situation of Nazir a khil'at and the titles of Nuwab Nazir and Khan to render his post respectable, such as the Nazirs attached to other Deorees have the titles of Nawab Nazir and Khan.[49]

Karl Wittfogel characterized the institution of eunuchs as a formidable weapon of autocracy for supervising and controlling the office-holding nobility. Hopkins and Patterson extend this further by describing it as a weapon in an absolute monarch's control and neutralization of the aristocratic classes. Such functions may have been critical in the Mughal and Awadh administrations, where the bureaucracy was noticeably constituted by 'Hindus'. The most obvious difference between Awadh and the Nazims in Bengal was that since the provincial Diwani had passed to the East India Company, and independent military commands followed suit, slave-deployment came to devolve more and more in offices of the household administration, and private armies and guards. This trend can be seen in the office of the Diwan Nizamat, an important office in the fiscal administration of the

[47] Letter of Nasir-ul Mulk, n.d., *MNLR*, I, pp. 22–23.

[48] Enclosed in Edmonstone to Harington, 15 April 1795, ibid., pp. 18–19.

[49] Ra'isunnisa to AGG, 7th Ramzan 1256, *MNLI*, II, p. 289.

household-state and which became a source of contention because of its inheritability. The eunuchs managing the qalamdānkhana in each de'orhi had greater power than the Diwan Nizamat. They were best qualified for this trust as they had no separate identity from their masters. As surrogates of the master, they could deal in financial spheres which were closest to the master's interest, yet retain within his/her control the office itself by virtue of their genealogical isolation as slaves. Hence early in the nineteenth century, an English official wrote every

> dependent, male or female, is subservient to them... through them the nominal Diwan receives what are called the Begum's orders, without the permission of the eunuchs he is not admitted to an audience, and then invariably in their presence.[50]

Clearly representative of the anti-hereditary principle, the eunuchs appear to have acted as counter-weights to the Diwan. Only once in the history of the administration was the power of the Diwan actually consolidated with that of the Nazim's eunuch: in 1852, with the dismissal of Diwan Sitanath Bose, an elderly eunuch Dārāb Ali Khan was appointed to the office.[51]

However, such explanations of political deployment of eunuchs also imply that males alone purchased or deployed eunuch-slaves. This was certainly not the case in many Indian ruling households, where older and powerful women purchased or held their own eunuch-slaves and deployed them to control both the rulers and establish independent administrations. In Awadh, for instance, the notable eunuch Almās Ali Khan, son of a Jat cultivator in Punjab sold to the father of Bahu Begam, came in the bride's dowry when she married Shuja-ud-daula, and managed her estates in Gonda and Faizabad.[52] Her son, Asaf-ud-daula, reared from childhood alongside this eunuch called him 'Mamu' (maternal uncle) and upon his succession to the throne in 1775 gave him the management of the entire Doab, equivalent to the farm of 23.2 per cent of all the state's revenues. Similarly, in the Nizamat, many of the eunuchs were acquired by the begams independently and individually, and not through one centralized recruiting centre.

[50] Suptdt. of Nizamat Affairs to Sec. to Govt., Secret, 8 August 1812, BC F/4/426/10,408.

[51] AGG Torrens to Sec. GOB, 24 June 1852, *MNLR*, II, pp. 634–5.

[52] Barnett, *North India Between Empires*, p. 182.

So, in 1818 five out of the six eunuchs serving in one de'orhi had been 'purchased by the late Babbu Begam,' and then taken over by the Nazim after her death and reassigned to serve one of his consorts.[53] The young eunuch Ruhan purchased by Wālida Begam, was similarly handed over on her death, to her heir the Nazim Walajah in 1823, and 'borrowed' from him by the Nazir of another de'orhi.[54] Mitamura argues that in the light of ruler-rearing structures, and the correlation of power with seclusion, psychological influence over the monarch could only be exercised by those servitors who, like the ruler, were born and/or bred in the palace, knew his foibles and could colour his information.[55] Hence, in both the administrations of Awadh and of the Nizamat, eunuch-servitors on the part of matriarchs influenced the rulers where the mothers themselves did not show their hands. In other words the presence of eunuchs on the part of matriarchs made rulership itself a two-cornered affair.

In the Nizamat, where most Nazims were young boys at the time of their accession, the power of the matriarchs and of their eunuchs was obvious. In 1772–3, after the accession of a young Mubarak-ud-daula I, and during the regency of Manni Begam, the Regent's eunuch, I'tibār Ali Khan, proved to be much more powerful than either the Nazim or the Clavering-led faction within the Council. Amongst the complaints that the Nazim made against Manni Begam, one referred to the administrative monopoly established by her eunuch thus:

> that most of the ancient servants of his Household had been divested of their employments... to make room for a set of people that he knew nothing of, the creatures of the Begum's eunuch—Atawar Ally Cawn... *the Eunuch had instructed the servants not to suffer him to learn any thing by which he might make himself acquainted with business.*[56]

In turn, one of the men, the Naib Daroga of the Nazim's Household Treasury, who had been dismissed by I'tibār, then found employment

[53] Statement of monthly distribution of Sicca Rupees 3797–4 from the De'orhi of Wālida Begam, through Roz Afzūn Khan, Nazir, BPC, 13 February 1818, no. 35, BC F/4/618/15418.

[54] AGG Loch to Pers. Sec. 22 July 1824, *MNLI*, I, p. 296.

[55] Mitamura, *Chinese Eunuchs*.

[56] Note on Nizamat Affairs, Home Misc. 584, p. 150, emphasis in original.

with James Grant, an English accountant at Murshidabad, to whom he complained against Manni Begam.[57] Officers of the East India Company recognized this two-fold insertion of the eunuchs in the administration: as agents of the male rulers and as a channel of communication with the matriarchs and female regents. As an official on the spot put it, a superintending Begam employs 'no other organs than her eunuchs; through them she sees, hears and acts'.[58]

Where rulership was divided between junior males and senior females, the conflict between the two resulted in the bid to take over each other's eunuchs. This was especially critical during the internal struggles of the Nizamat. In 1813, for instance, at the decease of Manni Begam, the reigning Nazim submitted to the Agent at Murshidabad a will supposed to have been written by her, and carrying the seals of the eunuchs of her de'orhi, Basant Ali Khan, Bahār Ali Khan, Tahsīn and Bahar Afzūn. None would have discovered this subterfuge if these men had not subsequently confessed the manner in which their seals were appended to the document. According to one version, the Nazim addressed them in these words,

> 'Now you are all become mine, and those who are desirous of advancing their interests will affix their seals to this will.'—Accordingly we... seeing that we had no protector, with a view to preserve our honour, in the spirit of obedience placed our seals before His Highness, saying 'Your Highness is our Lord and Master'.

In another version, separately issued under the seals of the eunuchs Tahsīn and Bahar Afzūn, it was said that

> four Hubshees were stationed over us for the purpose of obliging us to repair with the Seals to His Highness's presence. As it is human nature to have a regard for one's life and honour, we at last, having no alternative, repaired to His Highness's presence....[59]

Such attempts also meant that both sets of eunuchs could manipulate the ever-present hostility between the Nazims and the Begams. The officials of the Company utilized this breach in 1818 to garner

[57] Ibid., p. 147.

[58] Suptdt. of Nizamat Affairs, T. Brooke, to Sec. of Govt., 12 April 1813, BPC, 6 July 1816, no. 46.

[59] *MNLR*, II, p. 104.

information about the wealth and provisioning of the de'orhis. The eunuch Roz Afzūn Khan had been the Darogha of Babbu Begam's de'orhi for eighteen years, and in charge of the *Khwajaserai* from 1802. Upon the death of Babbu Begam in 1809, and the succession of the Wālida Begam, he was allowed to retire on his usual allowance of 200 rupees, Ambar Ali Khan, another African eunuch, was appointed to his post. However, in 1810, with the accession of a new, young Nazim, trouble erupted. The Wālida Begam, found the counsel offered by the Nazir eunuch, Ambar, to his young master fraught with danger to her authority. Accusing the eunuch of instilling 'pernicious counsel into the mind of the young Nabob in the same manner as it was notorious he had done earlier', she suspended him from the post, and reappointed Roz Afzūn Khan. When the latter was sent to the Nazim to 'present his nuzzer', he was refused, and threatened with the stoppage of his pay and customary allowances. This represented a humiliation meted out to his mistress, for her alleged breach of the system 'under which no Nazir or Ḍarogah of the deoree was appointed without the previous knowledge of the Nazim'. In 1817, therefore, when the Agent of the Governor-General was seeking information about the expenditures of the Mahalsarāi against the wishes of the Nazim, it was Roz Afzūn Khan he turned to: the latter 'filled with indignation and instantly resolved to expose His Highness's disingenuous conduct... furnished with an account of the appropriation of such portion of the sum... as was disbursed from the Walide Begam's Deurie'.[60] In the tussle between the Nazim Zainuddin Alijah, the head of the mahalsarāi Wālida Begam, and the representatives of the Company, the eunuchs proved to be a fairly important third force.

Given the nature of their individualized belonging, eunuchs became major targets of political manoeuvres and vendetta and acted as the scapegoats for the master or mistress. This was a consistent pattern both in conflicts within the Nizamat and in the relations between the Nizamat and the Company. Sometimes this extended to some eunuchs even turning upon each other. For instance Nazim, Walajah's eunuch cast aspersions on the 'proper eunuch' status (completeness of castration) of the head Begam's Nazir, Zamurrud.[61]

[60] AAGG Monckton to Sec. to Govt., C. Lushington, 26 December 1817, BC F/4/618/15418.

[61] AGG Magniac to Dy. Pers. Sec., 6 December 1823, BPC, 22 December 1823, no. 56 and from Nawab Nazim, recd. 18 December 1823, ibid., no. 58.

The transfer of slaves from one master to another, and the susceptibility of even powerful slaves to other slaves acting on behalf of another master is clear from the accounts. Together they suggest that stratification and differentiation among slaves was an important aspect of the organization of the Nizamat. As in the case of the Ottomans, who balanced African with Caucasian eunuchs, the Nizamat also seems to have acquired its eunuchs from different regions. The prefix 'Siddi' indicated that some of them were from Africa. But there were eunuchs of Indian origins as well, for instance Basant Ali Khan who was described in one record as a 'native of Hindostan... purchased together with another eunuch named Bahār Ali Khan... by Eithbar [sic] Aly Khan' and as 'a slave bought for the Begam when he was twelve years old'.[62] The eunuch Roz Afzūn was also described as a 'native of Upper Hindostan', purchased in his childhood by Babbu Begam: after the latter's death he was taken into the Wālida Begam's service, while Zamurrud, another eunuch, was acquired from Hyderabad by Basant Ali Khan.[63]

It is interesting to speculate whether the differences in ages, areas of origin, methods of individual acquisition and post-mortem transfers acted to keep the eunuchs from forming one corporate identity. The paucity of evidence at the moment prevents any but a superficial answer.

Evidence shows that stratification existed among and between eunuchs as among other male slaves and eunuchs. The eunuchs in turn had slaves and young trainee eunuchs. It is not quite certain whether it was the privilege of the Chief eunuch of the Nazim, or whether it was one shared by other eunuchs as well, to have his own slave, sometimes an entire establishment of slaves. For instance, Basant Ali Khan the eunuch commanded the services (and also drew the wages of) eighty one men and women, of whom eight were listed as khawās, at monthly rates varying from one to five rupees.[64] However two other lists exist, made up exclusively of chelas, one with twelve

[62] 'Statement in detail of Her Highness the Munni Begum's Monthly Establishment of Servants and Pensioners', by T. Brooke, BPC, 6 July 1816, no. 51.

[63] AGG to Pers. Sec., 6 December 1823, BPC, 9 January 1824, no. 56.

[64] Enclosure A, B, and D in J. Caulfeild to H. T. Prinsep, Sec. to Govt., Poll. Dept., 5 June 1838, *MNLI*, II, pp. 185–9.

names of those attached to the *Qadamsharīf* at monthly receipts of two to four rupees, and another showing thirteen names, without specific duties, but receiving between two and six rupees per month. All three lists pertain to the year 1239 (1833); in the last however there are some payments listed to the mothers and 'widows' of chelas like 'Mussamut Bodhun widow of a chelah whose name is not recorded' and 'Jan Beebee, widow of Seedee Punchoo Chelah'. From this we may possibly infer that these latter slaves or chelas were not eunuchs themselves, had been married, and served their eunuch masters. In addition, they did not enjoy the same privileges of rank that the eunuchs might have enjoyed. For instance, in 1853 while on a *shikār* expedition of the Nazim and the Agent, a tin box containing property worth Rs 700 was stolen from the custody of Hossainee Shaik, chela of the eunuch Miyān Arjumand, who though not a chief eunuch, had his own slaves.[65] The youth accused of having stolen the box, though described as a 'chelah of the Huzoor' did not belong to this group of eunuchs: his status was that of 'the son of a Gholam in the service of the Nizamat'. One can infer that this was a lesser status, compared to that of the eunuchs, because of the sleeping arrangements mentioned by witnesses; Etwaree, the boy's father, was described as living in a *pal* 'two or more haths from the Meeahs' tent, more or less, according to the nature of the ground'.[66]

The criterion operative in the formation of this hierarchy is not very clear. Going by the arrangement of space and functions within the Camp of 1853, one could infer that the highest rank among the eunuchs came from proximity to the Nazim. For instance, of the three principal tents one was reserved for the Nazim, one for Amān Ali Khan, an Abyssinian eunuch who was 'chief in dignity' among all the eunuchs, and the remaining tent for all the other eunuchs. According to an English lawyer, Amān Ali Khan's seniority was indicated by the post he held: that of Naib, 'or Lieutenant of His Highness. Under him, in various gradations were the others, including the Darogah, or

[65] Evidence of Hossainee, in Report of the Sessions Judge of Murshidabad, *Indian Records with a Commercial View of the Relations between the British Government and the Nawabs Nazim of Bengal Behar and Orissa* (1870, reprint Delhi, 1985), p. 171.

[66] Evidence of Mohammad Ameen, another eunuch in the service of the Nawab, ibid., p. 178.

superintendent of the Elephants, and of the kitchen department, also the Urzbegy (who is not a Meah)'.[67] Similarly, while Amān Ali Khan accompanied the Nazim on the hunt, the others did not. Conversely, though the other eunuchs 'regularly dined with the Nazim', the chief eunuch was permitted from time to time to absent himself from this attendance. Miyān Afrīn, a younger eunuch of the de'orhi of Begam Sahib, and not an eunuch of the Nazim, did not have the same privileges as did Amān Ali.

Length of service was an important determinant of ranking and stipends among the eunuch corps. In a list of Manni Begam's establishment of eunuchs, five junior African slave eunuchs appear to have received smaller stipends and allowances (of Rs 10 each) than the older eunuchs apparently 'hired' from other de'orhis (who received between Rs 25 to Rs 40 rupees each).[68] From the scattered evidence we have at present, the period that each had served is difficult to calculate. I'tibār Ali Khan, an eunuch in the service of the Najībunnisā Begam was said to have served thirty-four years in the Nizamat by 1836. Rough inferences can be attempted from the fact that the Nazim referred to the eunuch Amān Ali Khan as 'an old officer of my establishment, and the one of all others whom I had been taught from my earliest childhood to regard with entire confidence'.[69] In this instance, the age of acquisition or length of service can be calculated: Amān Ali Khan, in 1853, was stated to be 33, while the Nazim described himself as 25. Therefore this African eunuch had served by 1853, almost the entire period of the Nazim's life. In fact, by 1854 when he was dismissed by the Agent and the Diwan Nizamat, Prosunno Narain Deb, he claimed to have served the Nawab for twenty-three years.[70] Thus while Amān Ali Khan's seniority in position may have derived from this, it may be possible to infer from the ages shown against the other eunuchs like Arjumand (in 1853, aged thirty), Balāl (twenty-seven), Iqbal (twenty-five) that they would have served a smaller number of years, as compared to Amān Ali. However, seniority in age

[67] In the Court of the Sessions Judge of Murshidabad, 7 September 1853, ibid., p. 147.

[68] Statement of the De'orhi of Manni Begam, BPC, 6 July 1816, no. 51.

[69] Nawab Nazim to Duke of Argyll, Sec. of State for India, July 1869, *Indian Records*, p. 303.

[70] Petition of Aman Ali Khan, *MNLR* I, pp. 441–2.

itself could not have been the only criterion, since another eunuch in the same group was Jawahar Ali Khan, the sixty- year-old Nazir of the Bahu Begam[71], or even Haji Tamash who was fifty.

Within a literate, numerate bureaucracy, the power of eunuchs was enhanced by their function as administrative managers. However, only two references to their training exist. The first comes from the 1772 accounts of the household of Mubarak-ud-daula, in which we find the payment of tutors for the 'education of the slaves of the Nazim'. The second is a reference to the appointment of a tutor receiving a monthly wage of Rs 80 to 100, because the Nazim wished 'that the two eunuchs Meeans Uman and Nuzeer, his personal attendants, might be taught the English language,... they have themselves evinced a desire to learn'.[72]

Nevertheless, their literacy, account-keeping and proximity to their individual masters/mistresses did not exempt them from complete accountability to the individual whom he served. Thus for instance, Zamurrud Ali Khan, the eunuch Nazir serving Amīrunnisa Begam, was described by her as having 'incurred my displeasure in consequence of disobeying my orders by which I required him to furnish certain accounts, and he was then accordingly removed from his situation'.[73] Only a full settlement of accounts and express contrition averted further measures against him. This sequence of events was repeated again in 1852, when Nazar Ali Khan incurred the wrath of the Nazim and the Agent of the Governor-General for not having delivered a full set of accounts, for which reason he was accused of embezzlement.[74]

[71] The age of this eunuch appears doubtful, partly because in a list of Bahu Begam's de'orhi, submitted four years later, he is shown to have been seventy-five years old, vide Nawab Nazim to AGG, 22 September 1856, enc. in Extract Progs. Gen. (Poll.), 2 April 1857, BC F/4/2708/194267. The later list is probably inaccurate.

[72] Capt. G.D. Showers, Superintendent of the Education of the Nawab Nazim to Maj. Gen. F.V. Raper, AGG, 19 April 1843, *MNLI*, II, p. 463; Petition of Tamizuddin, late *atāliq* of the Nizamat College, to AGG Raper, 21 July 1845, *MNLR*, II, p. 552. W. Adam's 'Third Report on the State of Education in Bengal', 1838, is silent about the education of the sixty-three eunuchs in Murshidabad Palace.

[73] Amīrunnisa Begam to Lord Bentinck, 31 August 1833, BC F/4/1522/60090A.

[74] See correspondence between Offg. Magt., Murshidabad, C.F. Carnac, the AGG, H. Torrens, and Sec. to GOB, J.P. Grant, between February-June 1852 in BC F/4/2491/141010.

This represented the paradox of eunuchism itself. Important as an institution of anti-kinship, eunuchs individually remained susceptible to the conflicts amongst kin. By virtue of acting not just as the human barriers between the person of the master/mistress and the rest of the world, but also as his/her agents, their proximity to their masters and mistresses, and lack of wider kin ties was the pre-condition of their power. Yet they remained vulnerable, as slaves, to the wishes of their immediate masters, and often to the wishes of his/her heirs as well. A further paradox was that though individually unable to produce heirs of their own bodies, even where they were sometimes married to female slaves, they did reproduce subordination by acquiring younger cadet slaves and eunuchs themselves. It is this last feature that can be seen as paradigmatic of slave-reproduction as a whole in the Nizamat and elsewhere, principally in the pattern of recruitment of young female slaves by older skilled slave-women and freedpersons.

The World of Women: The Haram and Its Hierarchies

The head of the haram played a distinct role in governance. This position carried enormous influence, evident from the contemporary account of the dowager Begam of Murshid Quli Khan persuading her grandson to defer his accession to the Subahdari of Bengal,[75] and in the story of Alivardi's nomination to the Governorship of Bihar in 1732–3. Shuja, then Subahdar of Bengal, had proposed

> the appointment of his son, Sarfaraz, as the Deputy Governor of Bihar, [but] was opposed by his consort, Zebunnisa, who did not like to remain separated from her only son. She did not also support the appointment of Taqi Khan,.... She summoned Alivardi before the gate of the female apartment, had a rich khelat (dress) put on his body through her son Sarfaraz, and gave him a firman for the Government of Bihar. After this investiture Alivardi was called before Shujauddin, who also on his own part gave him an elephant, a sword, a dagger, an embroidered head-dress along with other presents, and the patent for the Deputy Governorship of Bihar.[76]

[75] Ghulam Hussain Salim, *Riyaz-us-Salatin* (trans. A. Salam, reprint 1975), p. 288.
[76] K.K. Datta, *Alivardi and His Times* (Calcutta, 1963), p. 9.

Thus the governor was *first* selected by Zebunnisa, and confirmed by the reigning Subahdar. Similarly, Alivardi also reaffirmed this position of the head in his turn. After the usurpation he is reported, to have solicited the pardon of Sarfaraz's sister, Nafisa Begam, *before* ascending the masnad.[77]

Like the Nazim who succeeded to the masnad, the Head of the Mehal who succeeded to the *gaddi* gave out titles and honours and above all, the allowances and customary presents. The rituals of investiture of the Nazim and the head of the Mahalsarāi were similar. Though we have accounts of the former, very few records exist of the latter's investiture, except that of the Wālida Begam (Faizunnisā) when she moved to the apartments of the Babbu Begam, when 'several of the branches of the family... attended her in the procession and presented Nuzzers'.[78] Since the ability to sponsor public buildings was another sign of authority, Manni Begam's construction of a *masjid* and a *naubatkhana* for her son were important in indicating the funds and labour she could command. A later official commented that

> the Musjid, to which on festivals and days of ceremony the Nazims of Bengal were accustomed formerly to resort, having been destroyed by the encroachments of the River, the Munni Begum resolved to build another at her own expense in the Chowk which she had established in the city; and as the beating of the Noubut formed a part of the honours which were usually paid to the Nazim on these occasions, she also constructed a Naubut Khanah in the vicinity of the Musjid.[79]

From Reza Khan's account, it appears that besides the publicly acknowledged position of the head of the Mahalsarāi, the power of the *gaddinashin* begam was reckoned in a multitude of ways.[80] Not only did she have the power to invest significant individuals with the khil'at, but the position was also associated with substantial fiscal privileges, like the income from the *ganjs* which, originally held by

[77] Ghulam Husain Salim, *Riyazu-s-Salatin*, pp. 321–2.

[78] R. Rocke to J. Monckton, 8 August 1811 in BC F/4/372/9260.

[79] Extract Poll. Letter from Bengal, 17 March 1820, BC F/4/732/19792.

[80] Naib Subah's letter in *Proceedings of Controlling Council of Revenue at Murshidabad* (ed. W.K. Firminger, Calcutta, 1919), under date 3 Jan. 1771, III, p. 10.

'the mother of Mīran',[81] had descended to Manni as the mother of the Nazim Najm-ud-daula. This had continued in her possession subsequently as well. When Mubarak-ud-daula became Nazim, therefore, he sought to transfer these ganj, namely Jafar Ganj, Maklama Ganj and the Chauk,[82] to his mother.

The privilege and authority that was associated with the position of the head of the Mahalsarāi was strengthened when the ruling Nazim was her son. So critical was this power that even when Najm-ud-daula died and Saif-ud-daula came to the throne, Manni Begam retained the gaddi, because as Reza Khan was to put it later, the latter had no mother.[83] But this changed in 1770, when Mubarak-ud-daula ascended the masnad. Born of Babbu Begum, this meant that the gaddi should have gone to his natural mother. In the words of Reza Khan, formerly

> the most respected of all Myr Jaffiers Begum... was the mother of Sedek Ally Cawn [Sadiq Ali Khan, Mīran]. After the death of Sedek Ally Cawn when as Nigim & Doulah [Najm-ud-daula] was the next son; His mother rose to Preminence after Myr Jaffiers Death, when the Nizamet devolved to Nigimul Dowlah his mother derived her claims from Her Sons Station as Nazim.[84]

In the conflict that erupted, Reza Khan proposed that authority and rank be shared by both, though the 'right is in the mother of Mebareck ul Dowlah'. For Cartier, this was impossible. So he invested Babbu Begam in the gaddi. This, however, did not resolve the conflict between the two Begams. As this correspondence reveals, the central locus of

[81] Shah Khanam, step-sister of Alivardi Khan, who had born to her one son, Sadiq Ali Khan (Miran) and a daughter who was married to Mir Qasim. Shah Khanam died 'in the time of Nidjim-ut-Dowlah' according to Reza Khan, which presumably indicates sometime in 1766.

[82] From a list of 1790, it appears that the ganj in question were transferred to the holding of Babbu Begam. From thirty-eight shops in these three ganj which sold liquor, a revenue of Rs 2205 accrued to the holder, *See* Collr. J.E. Harington, 16 August 1790, BBOR (Misc.), 29 October 1790, Statements 1–6.

[83] Historians have, uncritically in the past, described Saif-ud-daula as a 'son ' of Manni Begam without differentiating between the two kinds of sonship and motherhood that were contested in the Nizamat. This is contrary to the description that Muhammad Reza Khan gave of Saif-ud-daula as 'having no mother' in a letter to the Council at Murshidabad, *Progs.*, 3 January 1771.

[84] Letter of Mahomed Reza Cawn, recd. 26 May, 1770 in Home Misc. 202, p. 60.

power in the Mahalsarāi was the figure of the mother of the Nazim, not the wife. The importance of this identity, that of mother, was attested to in 1770 when, during the conflict between Manni Begam and Babbu Begam, the latter's supporter reiterated the distinction of status that separated the childless woman from the mother of a son and heir thus,

The Dependants of Mahomed Jaffer Chan [Khan] are all 12 in number including Manny Bhegam and the Mother of Mobarek-ut-Dowlah, the remaining 10 excepting one who has got a Daughter are all without Children. From the Time the Mother of Mobarek-ut-Dowlah, agreeably to the Orders of the Council, became the principal in the Mahalsarāi She has supplied all these dependants and their Slaves, with Expenses, Provisions, Cloaths and every other necessary without having the least Relation to Munny Bhegum. In like manner She has the Care over 12 Women of the Nabob Syef-ut-Doulah, but as the Nabob Nudjim-ut-Dowlah was born of Manney Bhegam, she has the Charge of 4 of his Dependants whom she supplies with provisions, Cloaths and Necessaries from the Nabob Mobarek-ut-Dowlah.[85]

It was not that Manni Begam contested the right of a mother to the gaddi, but that she contested Reza Khan's notion of motherhood. She claimed that Mīr Ja'far,

according to custom, committed him to my Bosom,& having brought him up for 12 Years, the Nabob Mozuffer Jung has now separated him from Me... he has to my Dishonor publish'd that Bew the mother of Mobarik ul Doulah is appointed to the principal authority in the Mehal Serai,....[86]

Two differing conceptions of motherhood seemed to be at issue: Manni claiming the position of chiefship not by virtue of a biological mandate, but by virtue of having nurtured Mubarak-ud-daula ('for 12 years') and Reza Khan, conceiving of chiefship of the Mahalsarāi strictly in terms of the blood-tie with the Nazim. Which concept of motherhood, and therefore the succession to the headship of the de'orhis,[87] was to prevail depended on a host of factors, not the least being the Company.

[85] Letter from Mahomed Reza Khan in *Proceedings* (ed. Firminger), III, p. 10.

[86] Arzee from Munny in *Proceedings* (ed. Firminger), II, pp. 95–6.

[87] *De'orhi*—literally threshold—came to mean a particular set of apartments in which a matriarch, her sons, daughters and other dependents resided.

Company-made Heads

In 1772, Warren Hastings' restoration of Manni to the principal position in the Mahalsarāi was tied up with the commercial interests of the East India Company, whose representatives from 1770 had been pressing the Naib Nazim, Reza Khan, to curtail the expenditure of the Nizamat and divert the funds towards maintaining some battalions of English troops. Ousting Reza Khan, who had championed a genealogical mother's claim, and appointing Manni Begam to the Regency was significant because the latter had been privileged on grounds of her current *childlessness*:

> The Begum... has no children to provide for, or mislead her fidelity. Her actual authority rests on the Nabob's life, and therefore cannot endanger it: it must cease with his minority, when she must depend absolutely on the Company for support against her ward and pupil, who will then become her master.[88]

Henceforth, the headship of the mahalsarāi would only go to women who were childless at the time of their elevation, justifications of the Company be what they may. For instance, after the decease of Manni Begam in 1812, the Council at Calcutta insisted upon the elevation of the 'seniormost' woman, Faizunnisā, one of the concubines of Mubarak-ud-daula I, to the gaddi. In its instructions to the Superintendent, the Government observed

> that the age, rank and character of the Waulida Begum would alone render her the fittest person to have the guardianship of the late Munni Begum's Deurie... by the usages of the country and the actual practice in the Nizamut it becomes a positive right, of which she could not be justly deprived.[89]

This emphasis the Company put on the 'advanced age' of the woman was not the whole truth. Age had not been the critical factor in Manni Begam's or Babbu Begam's claim to authority, as much as her rank as mother of a son.

Every Nazim therefore had to contend with a chain of command made up of grandmothers and stepmothers and not any woman who

[88] Fifth Report, Committee of Secrecy, 1781, Appendix, IV cited in James Mill *The History of British India* (fifth ed., revd. H.H. Wilson, London, 1858), III, fn., p. 379.

[89] Monckton to Brooke, 8 February 1813, *MNLR*, II, pp. 101–4.

was in the category of 'wife'. After Hastings' departure in 1784, a more consistent attempt was made to draw the Nazim away from the sanctuary of the haram, and the Begams. The actual physical relocation of the Nazims, away from the Killah, began with the plans of 1802 to build a new palace. By 1808, the Nazim Babar Ali had been induced to move into Nishat bagh, four miles from the Killah,[90] a removal that deeply offended Manni Begam.[91] The investment by the Governors-General and the Company's government, in the building of a new Palace from the 1820s, had small beginnings but significant political implications.

The relocation of the Nazim, and the building of a new palace under the aegis of the Company sharpened the conflict between various Nazims and Begams over the size, embellishments and location of accomodation of each woman, and the display of manpower at each.[92] Thus when Babbu Begam died, that authority went to the mother of the incumbent Nazim, Faizun (Wālida) begam. In 1811, with the accession of another Nazim, Alijah, she was asked to vacate the apartments in which she had been installed by her son, and relinquish the prerogative of distributing 'victuals, clothing and other customary rights and privileges'.[93] The significance of the apartments was emphasized by Zainuddin; the de'orhi of Babbu Begam, he urged, should not only be maintained 'in a more splendid style and on a larger scale than the others', but that it should be occupied by his consort and himself.[94]

In effect a fundamental shift occurred when the Government authorized an allowance of Rs 300 to the Nazim's first consort and allowed her to occupy the apartments vacated by Wālida Begam. The authority that Wālida Begam had hitherto exercised by virtue of her

[90] Pattle's private note to Governor-General Minto, Minute of 28 January 1808, BPC, 1 February 1808 no. 65.

[91] Note to Pattle, recd. 17 May 1810, saying that Babar Ali over 'the last two years and a half... acted in opposition to my will... removed from the Killah which was near my residence and went to live at Nishat Bagh and Furrukh Bagh', BC F/4/312/7144.

[92] Suptdt. of Nizamat Affairs, R. Rocke, to J. Monckton, Pers. Sec. to GOB, 31 January 1811, BC F/4/372/9260.

[93] Letter from Walida Begam, received 7 February 1811, ibid.

[94] Translated letter from Nawab Nazim, to Mr. Rocke, n.d., and J. Monckton to R. Rocke, 28 June 1811, ibid.

eldership and her motherhood of a Nazim was compromised. The Earl of Moira, who visited Murshidabad shortly after this, perceived this shift as a difference in styles of decoration of each of the women's apartments he visited. Taken first to the Wālida Begam, whose abode just behind the grand hall of audience was marked off by 'sheets or coarse table-cloths sewed together', he implied that a breach of the ranking system had occurred since another woman, the 'favourite' wife of the Nawab Zainuddin, was 'better lodged than the Walideh;.... The curtain was of rather better materials'.[95]

Headship was achieved through motherhood, and one elder's death caused the promotion of another. But the Walida's grandson's (Nazim) wishes to initiate a rule by youth over the traditional rule by elders in the haram created a space for the redefinition of the criteria of headship in the hands of the Government in Bengal.

In 1818, the Government sanctioned one of the Nazim's consorts a personal stipend of Rs 2000 per month, which had no precedent. It was a reward, for the Nazim's 'commendable' conduct in agreeing to 'discharge the floating debt out of the late Munni Begum's property', agreeing to reduce his household to the amount of Rs 14,400, thereby relieving a financially straitened Company of a greater portion of its obligations.[96] Around 1821, just when the succession to the masnad in the outer court was reverting back to a brother, the principles of succession to the inner government began to change at the initiative of the Magistrate of Murshidabad, Magniac. Instead of motherhood to the Nazim, other criteria, like the form of marriage, birth, the degrees of affinity as the British understood it began to be brought to bear upon the choice of the candidate. Thus Amīrunnisa's claim to the position began to be privileged on the grounds that she was 'the first cousin and favourite wife' of Zainuddin: moreover, it was claimed that 'she had been married to him by the Shadee Ceremony'.[97] This was a new concern: few officials had raised this issue when promoting the claims of Faizunnisā' to the headship earlier. Furthermore, there were those women who were older, like Sharīfunnisā', another of the

[95] Hastings, *The Private Journal,* I, pp. 83–4.

[96] Poll. Letter from Bengal, 4 June 1818, paras 58–90, BC F/4/372/9260.

[97] Minute and Resolutions of the Governor-General in Council, 20 March 1822, *MNLR,* I, pp. 207–11.

concubines of Mubarak-ud-daula I, also the mother of Mīr Manglī, apart from the first-acquired consort of the Nazim Zainuddin Alijah, referred to as the Bahu Begam. The claims of the former were dismissed on the grounds that she was 'only a khanum' and those of the latter on the grounds that she was 'a mere Khuwas, of no family or respectability; and His late Highness never manifested any desire to bring her forward into any prominent situation of the kind'.[98] Yet, Government was apprehensive of recognizing Amīrunnisa's claims to the headship immediately as her father, Akbar Ali Khan Shamsher Jang, held an influential position in the court of her husband, Zainuddin. Thus it was not till the accession of Ahmed Ali Khan Walahjah, Zainuddin's brother, that Amīrunnisa was recognized as the Head of the Mahalsarāi herself. Amīrun's elevation by the Company, however, completely ignored their *own* criterion of age-based seniority, since Amīrun was in her late teens and younger than the Nazim Walajah..[99] Moreover, she was no Nazim's mother.

Such manipulation of the headship was often employed by English agents and officers. For instance, when Najībunnisā, mother of the Nazim Humayunjah died in 1860, the Under Secretary to the Government of Bengal wrote that 'these *de'orhis* are intended for widows and not for wives;... further restricted to those widows who have held *the first rank as wives*... the number of de'orhi-holders is uncertain and must depend entirely upon the number of chief wives who have survived'.[100] Motherhood was completely ignored.

This created a sharp rupture in the structures of authority within the de'orhis. Since Amīrun continued to be the head till her death in 1858, her tenure coincided with two changes of Nazims in the outer mahal. This meant that instead of the *mother* of each Nazim receiving the highest stipend, and the control of other de'orhis, Amīrun continued to be supreme. In 1840, she received the highest personal allowance of Rs 7326, while Najībunnisā the mother of a deceased Nazim received Rs 5568 and Ra'isunnisā, the mother of the ruling

[98] Ibid.

[99] Note from the Nawab Nazim (Walajah), recd. 18 December 1823, refers to Amīrun as 'my cousin and my junior in years', BPC, 8 January 1824, no. 58.

[100] H. Bell cited in Ch. Sec. GOB to Sec. GOI, Foreign, 11 July 1893, WBSA, Judcl., November 1893, no. 57.

Nazim, received only Rs 2749.[101] Amīrun's predominance till 1858 spelt the escalation in conflict between elder and younger within the haram. Speaking generally of the relations between the other 'Ladies of the de'orhis' at the time of Amīrun's ascent, an official noticed that they

> jealous of the elevation of the Doolhine Begum and annoyed at being under the control of one much younger than themselves and who, when the favorie lady of the Nazim, treated them with such harshness or neglect, are not cordial.[102]

Mothers of successive Nazims vied for equivalence with Amīrunnisa, at the same time that the consorts of each deceased Nazim tried to meet the English government's insistence on 'chief wife' status by claiming it for themselves. Yet both failed in the face of a colonial government determined to cut costs, retrench posts, and undermine the mobility within these structures.

Frozen over: The Hierarchy of Motherhood, Service-skills and Seniority

As the Company sought to undermine the political role of the haram by manipulating its headship, it also simultaneously devalued the importance of motherhood in determining the potential movement of some female slaves inwards and upwards into positions of privilege. Till Amīrun's headship, the power of the chain of matriarchs in the zanana, was remarkable because of the fact that the political position of mother had had little to do with previous jural status. Slaves like Manni Begam, and the slave-born like Babbu Begam, through the service of either bearing children, or through rearing them, could eventually assume positions of command. Manni Begam was born to a widow of Balkunda, a village near Sikandra; she was sold to Bisu, Sammen Ali Khan's slave-girl, who lived 'for five years at Delhi where she taught Munni the art of dancing'.[103] Brought to Murshidabad

[101] AAGG Raper to Sec. GOB, 20 January 1841, *MNLI*, II, pp. 286–92.

[102] AGG to Pers. Sec., n.d. BPC, 2 May 1823, no. 88.

[103] *See* Brajendranath Banerji, 'The Mother of the Company', *BPP*, 32, 63–4, 1926, pp. 38–48; and *Banglar Begum*, (Calcutta, 1913); P.N. Bhalla, 'The Mother of the Company', *Journal of Indian History* (henceforth *JIH*), 22, 2–3, 1943, pp. 128–44.

around 1745–6 by Bisu who had been summoned by Nawab Shahamat Jang (Nawazish Muhammad Khan) on the marriage of his 'adopted' son Ikram-ud-daula—the younger brother of Siraj-ud-daula—the whole party seems to have been subsequently taken into the service of Mīr Ja'far at a monthly stipend of Rs 500. This group also perhaps included the daughter born of Sammen Ali Khan to Bisu, known subsequently as Babbu Begam, the second concubine of Ja'far referred to by Vansittart. This was repeated in the lower echelons of the haram. Slave-girls, trained in various skills including music and dance, appeared simultaneously as concubines of a son and servant of some matriarch. Sometimes, as in the case of Faizun, and Sahib Jan/Ra'īsunnisā, they eventually became de'orhi heads themselves.

From the details of the hierarchy available from 1816–18, we can get a fair idea of the internal economy of the de'orhis, which enables us to understand the hierarchy of powers and functions that prevailed in these segregated domains. This section will analyse the composition of the de'orhis which were under the control of the Wālida and the Bahu Begam, that is the Raunaq Afza and the Choughura de'orhis.[104] Raunaq Afza contained the concubines and relatives of previous Subehdars and Nazims, including those of Mīr Ja'far, and the Choughura de'orhi were the apartments of the concubines of Saif-ud-daula.

There were broadly three groups in the establishment paid by the Wālida Begam. The first comprised the children and the concubines of previous Nazims, whose personal stipends varied enormously. An age-based, or generationally descending scale, seems to have applied here, concubines of fathers receiving more than concubines of sons. Thus the concubines of Mīr Ja'far, like Bee Bugloo, received Rs 80 per month, and those of Saif-ud-daula received a personal stipend of Rs 65 each. Also, they were distinguished according to the differing kinds of motherhood they had achieved. Hence, one of the two concubines of the Nazim, Zainuddin Alijah, who was also the mother of his eldest daughter, received Rs 50 monthly, but the concubines of his predecessor, who were not mothers of his children, received stipends ranging from Rs 15 to Rs 20 monthly. This was equally true of the

[104] The data for 1818 is taken from Monckton's survey in BPC, 13 February 1818, nos 33–42, BC F/4/618/15418.

concubines of Mubarak-ud-daula I: those women who had borne children uniformly received between Rs 150 and 200 monthly, like Sharīfunnisa Khanum (the mother of Mīr Manglī, and Zībunnisa) and Mubarakunnisa (the mother of Mīr Muhammad Ali and Jigri Begam), whereas his other concubines received between Rs 20 and 30 monthly. Other determinants of rank like skill or length of service may have also determined the ways in which these stipends were given.

Apart from the group of mothers and concubines, the de'orhis maintained a group of service personnel which included *mughlānī* (waiting-women) trained in literary or accounting skills, as is implied by the description of one such mughlānī being employed as a 'Secretary'.[105] Other skills and services were those of *farrāsh* (care of carpets and tents), seamstresses, *huqqabardār* (preparer of the huqqa, for smoking), wet-nurses and mid-wives with the prefix 'dadda', 'mama' and 'dai'.[106] In a polity where becoming a mother was important, the availability of slave-midwives and nurses was critical for the survival of any infant. At the same time, having suckled an infant, the potential of becoming a 'foster-mother' to a Sahibzada or Sahibzadi and thus moving up the hierarchy was ever-present. One such slave-nurse was considered to be a major figure in the haram of the Nazim Walajah: children of these slave-nurses, counted as foster-siblings of the infants they had suckled. A particularly strict regimen was attached to such service. Wet-nurses were required to be abstemious in their diet. An observer in Lucknow noticed that they were

> not allowed for the first month or more to taste animal food, and even during the two years- the usual period of supporting infancy by this nourishment—the nurse lives by rule both in quality and quantity of such food only as may be deemed essential to the well-being of the child.[107]

[105] Statement in BPC, 6 July 1816, no. 51.

[106] The allowances of individual slave wet-nurses and midwives could vary according to the rise in power or status of the child she had suckled. Thus one of the sons of the Nazim Mubarakud-ud- daula I, Sayyid Ali Reza paid to 'Dada Noorun who nursed [him] in his infancy' a sum of Rs 16 per month, independent of the sum of Rs 12 that this woman received from Babbu Begam. *See* AGG to C. Lushington, 23 September 1817, BPC, 10 October 1817, no. 36.

[107] Mrs Meer Hassan Ali *Observations on the Mussulmans of India*, revd. and ed. W. Crooke (reprint Delhi, 1975), pp. 211–12.

A third group, called *asil,* in turn served the previous group. Some of these asīls may have performed those functions and duties attributed to slaves in other harams like story-telling or *champi* (the contemporary English term was shampooing) or massaging of limbs.[108]

In the Choughura de'orhi, the stipends of the service and skilled corps were higher, older mughlānīs received Rs 25, while their juniors got between Rs 5 and 10. The same held true for the *gāens* : the best-paid received a personal stipend ranging from Rs 20 to 25; a middle group received Rs 14–15 and the lowest-paid gāens received Rs 9 or 10. However, it is not easy to determine on what grounds these variations occurred, whether on the length of service, as in the case of the mughlānīs, or whether according to the degrees of skill. In addition, some gāens who received higher stipends like Jahānabadī, Nawab Bai, Nūr Jahan and Rattan Kunwar, did so as ('haram') concubines of a Nazim.[109] For the seven other gāens receiving personal stipends of Rs 10–15 each, no such additional description exists in the record of 1818. This suggests that even among a group of skilled female slaves, certain distinctions existed: perhaps physical attractiveness and age distinguished those who would become 'sexual' from those who did not, though this is a conjecture on my part. However, in this distinction among female slaves perhaps lay a clue to the mystery of the differences of service in the de'orhis: as some female slaves grew older, they may have been reassigned as servants of the begams, or of the Sahibzadas and Sahibzadis. Certainly, the overlapping of function and status characteristic of the Murshidabad mahalsarāi may be thus explained.

An alternative principle that distinguished servants from concubines may have lain in the methods of recruitment and training as specialists. For this reason, it is significant that three girls were attached to the gāens at the upper levels in the Choughura de'orhi. The junior's individual stipends of three rupees could have denoted either their age-based juniority or their inferior ranking among the

[108] Fanny Parks reported greater specialization among the female slave masseurs, one doing only the head and not the limbs, in *Wanderings of a Pilgrim in Search of the Picturesque During Four-and-Twenty Years in the East; with Revelations of Life in the Zenana* (Karachi, 1975), I, p. 450.

[109] Colonial officials used the term 'harem' to denote 'a secondary, or lower, position among the ladies of the Nawab's zenana', cf. Offg. AGG to Sec. GOB, 9 July 1874, NAI, FPP., August 1874, no. B5.

service group attached to the establishment as a whole. However, what is truly significant is that the three 'juniors' are called *chokree*, indicating a pattern of renewal by which slave women either acquired or bore other female slaves. A British official described the gāens paid by Manni Begam as 'girls purchased [who] were instructed in Singing—the Women on the present Establishment are advanced in years, but as they die away their places are filled by Girls born in the Deury'.[110]

The pattern of slave-concubines and skilled performers acquiring slaves, suggested by the records of 1818, is confirmed by a comparison with de'orhi lists of the 1850s. Thus from the de'orhi of one of the daughters of the Nazim Mubarak-ud-daula (Badrunnisā, also a wife of Shams-ud-daula, nawab of Dacca), the sixteen female khawās had served between twenty and fifty-eight years,[111] and had entered service at ages between six and fifteen. Comparing the 1854 list of the Choughura de'orhi with that of 1818 we get the impression that despite the Company's efforts at resuming the allowances of all who were not 'family' but only 'dependents', the patterns of recruitment into the de'orhi remained what they had previously been. Thus, in 1854, among the gāen of the Choughura, apart from Murād Bakhsh who was described as an 'aged woman... the only Gaen now alive' from the list of 1818, there were ten fresh names, to the despair of the Agent who could not fathom when these had been added to the records of the Sherista Nizamat.[112]

A significant feature of the allowances meted out by specific heads of de'orhis was that there was no automatic correspondence between the de'orhi from which their stipends were issued and the actual place of residence of the recipients. The payment of these stipends determined who possessed final control over them, and only secondarily their residence. This explains the division between the female khawās (slave-attendants) some of whom were paid by one Begam and some paid by another, though all of them seem to have

[110] Brooke's remarks on Enclosure F, Statement in BPC, 6 July 1816, no. 51.

[111] Statement of the old servants of the late Badrunnisā Begam in AGG, Lt. Col. Macgregor to Sec. to GOB, 10 February 1857, BC F/4/2708/194266. *See* Appendix I, Table II.

[112] Remarks by AGG on statement of present monthly distribution of De'orhīat, BC F/4/2688/187904. *See also* statement in Collr. Murshidabad to Commr. Presy. Divn., 13 July 1893, WBSA, Judcl., November 1893, encl., no. 90.

cquired during the reign of Mubarak-ud-daula I. The impression they may have been transferable from one de'orhi to another is engthened by the fact that some of the women of the lesser service orps, though paid by the Wālida Begam, resided in the apartments of Bahu Begam. Similarly, some concubines stated to have apartments in the Choughura appeared to have resided elsewhere. As explicitly stated by the steward of a son of Mubarak-ud-daula I, many of the slave-servants and concubines did not 'reside in his House... they received wages from the Bubboo begum's Deurie and he [Mīr Moghul, the son] made them occasional presents consisting of Articles of Wearing Apparel'.[113]

Focussing on actual residence helps us to understand that 'domestic' slavery did not imply that slaves of the household *only* resided *in* the household of the person who first acquired them. As we have seen with the eunuchs, a large number of female slaves, first acquired by the Babbu Begam, were found subsequently 'stationed' with the sons and daughters of subsequent Nazims, many of whom were themselves born of concubines.[114] Thus both in the cases of Mīr Moghul touched on above, and of Saiyid Ali Reza (aka Mīr Moulvi), both of whom were sons of Mubarak-ud-daula I, the concubines and slaves were either 'presented ' or 'transferred' by the matriarchs of the de'orhis. The distinctions between the servants and the served lay in the different stipends received by each group. For instance, female servants received between three and twelve rupees, while those they served, like the daughters of Mubarak-ud-daula received between Rs 250 and 500. Similarly, for the male slaves,.and those who were stationed with the Sahibzada, there was a vast gulf in stipends. Apart from the spatial mobility generated by the inheritance and gift transfers of slaves from

[113] AAGG Monckton to C. Lushington, 8 October 1817, BPC, 17 October 1817, no. 52. The steward Mirza Khadem Ali, detailed the origins of the five women on whose petition for subsistence this investigation had occurred. 'Bee Jaun was purchased by Meer Mogul for 200 Rupees,... Bee Nafurman was transferred to his service by Mubarak Khanum, the mother of one of Mubarakud Daulah's sons,... Goonbharree and Sook Chyne had been formerly slave Girls of his Mother and... Noorunissa had been his wet nurse.'

[114] For a detailed break-down of slaves paid from specific de'orhis but serving in others, *see* enclosure in Pers. Sec. to F. Magniac, 25 July 1822, BPC, 26 July 1822, no. 79.

one de'orhi to another, intra-generational mobility could sometimes translate into spatial distance between masters and slave-servants. As the example of a singing woman in her nineties, Laboo Bai, reveals, age and skill could remove some from direct physical proximity with, and control of, her masters: she had served Mīr Ja'far, and then Manni Begam for 58 years, for which she received Rs 200 and resided in the Chauk bazar in Murshidabad.[115] Age was a significant element of stratification among female slaves in other households as well. The semi-autobiographical narrative by Mīr Musharraf Hussain, *Udāsin Pathiker Moner Katha*, described this stratification among female slave-servants (*kharīdā*) in one substantial household of Bengal in the nineteenth century. Speaking of elderly female slaves, he wrote of one among them enjoying privileged stature by virtue of the fact that her daughter was one of four select personal attendants of the female holder. Additionally, the younger female slaves were distinguished from the elders by the practice of seclusion. Thus the four personal attendants (*khas bāndi*) Durgati, Harūn, Nūran and Champa never left the house, whereas Durgati's mother, the privileged elder female slave Sabja, enjoyed unrestricted mobility.[116]

The non-correspondence between residence and the source of a stipend helps explain the phenomenon of male slaves, comprehended within the group Shāgird Pesha, who, though paid by the Begams, did not reside within the mahalsarāi at all. Information on the specific residences of the large majority of slaves is unavailable except for the eunuchs on Manni Begam's establishment, who were said to 'have all separate houses and separate establishments'.[117] As for the non-eunuch male slaves, it seems likely that they performed certain important tasks *outside* the de'orhi as its agents, and may also have lived on lands and houses belonging to the de'orhis and the Nizamat. It is in this capacity that the individual khawās of the Shāgird Pesha received between three and five rupees each, whereas the men who tended the gardens in and around the city which belonged to the Nizamat appeared to have received even less. Higher stipends paid by the mahal

[115] Enclosure K, Statement in BPC, 6 July 1816, no. 51.

[116] Cited in Kazi Abdul Mannan (ed.) *Mosharraf Rachana Sambhar* (Dacca, 1976), I, pp. 578, 582–3. The narrative itself was first published in 1891.

[117] Statement by Brooke, BPC, 6 July 1816, no. 51.

to certain members of the 'Mulazimān Imtiyāzi'(literally distinguished servants) included men like the Darogha who managed the markets at Patna on behalf of Babbu Begam, and received Rs 40 from her. Within this group of higher servants of the mahal were also the Begam's principal *munshis, mohurrirs,* the *Khufiyanavis*, (a post equivalent to the secret intelligencer under the Mughals), the *Qāris* (readers of the Quran), and the chanters of the *fātīha* at tombs.

Recruitment of such servants was linked to the mahal's internal structure. Such men appear to have been the sons of the nurses and mughlānīs in the de'orhis. For example, one of the highest stipends issued by the Wālida Begam in this category—Rs 58 per month—went to a man who was described as a foster-brother of Mubarak-ud-daula I; he was born to the wet-nurse of the infant Nazim. Similarly, another who received Rs 10 per month, was the son of a favourite mughlānī of the Babbu Begam. Yet other male servants within this category of Imtiyāzi, whose stipends were paid by the Bahu Begam, were other such 'sons of wet-nurses or 'grandson' of a concubine. After the 1820s many slave-women who had achieved motherhood by some means or the other, began to find it emptied of both material and symbolic value. Officials denied the claims of all 'adopted children' of concubines upon the stipends and allowances that had been paid to their adoptive mothers. Alongside this recommendations to separate children and mothers became common. Nasīruddīn Haidar's concubine's daughter, Hussainī, was separated from her birth-mother, 'a Woman of Ill repute', and placed 'under charge' of the 'legal wife' though the latter was estranged from her husband.[118] The Government of Bengal though aware that such a measure might be objectionable, suggested that the 'family... find means of reconciling her to the surrender of the child'.[119] Another slave-concubine of a grand-uncle, mother of a three-year-old son when her master died, had her son taken away from her and sent to the newly established Nizamat School[120]: three years later, the mother too left Murshidabad 'with a person under the name of John Kranee' [*kerani* in Bengali or

[118] AGG Russell to Pers. Sec., 4 March 1822, BPC, 11 April 1822, no. 128.

[119] Pers. Sec. to AGG, 30 March 1822, ibid., no. 129.

[120] AGG Thoresby to Sec. GOB, 5 December 1833, Extract Poll. COns. 19 December 1833, no. 100, BC F/4/1475/57975.

writer-clerk] and her name was erased from the records of stipendaries.[121]

Cleaving the Polity Apart

The Company's fiscal policies, and the personal influence of the Agents, escalated the conflict between male kin and affines at the same time that they caused a divorce between the political role of the de'orhis and their social existence. In the first half of the nineteenth century, officers of the Company persisted in nominating younger males over elder collateral males to the office of Nawab Nazim. The Government of the Company then threw its might into extracting the ceremonial deference owed by these marginalized elder males to their younger kinsman. Individual AGGs intervened directly to exile recalcitrant male kin from Murshidabad periodically when their efforts to control them were infructuous.[122] Such proceedings created precedents: Nazim Humayunjah cited the orders of government of two decades previous to his that peons be sent to coerce the attendance of his older kin (one of whom was a grand-uncle).[123] This was particularly ironic since in the matter of the headship of the mahalsarāi, the Government had avowedly adopted the principle of seniority; and in the matter of appointment to administrative posts in the Nizamat, continued to avow principles of reciprocity and affective kinship.[124]

Similarly, while interceding on different sides in the disputes between matriarchs and Nazims, the Company also substantially modified the basis of authority in both and thus helped to diminish both. We can only outline the multiple shifts that occurred between 1775 and 1844. With the coming to age of Mubarak-ud-daula I, the political role of the haram came to be increasingly resented by the Company's officials because it indicated their failure to direct the actions (and expenditures) of the nazim. In 1787, for example, it was

[121] AGG to Sec. GOB, 2 December 1836, BPC, 6 December 1836, no. 12.

[122] For the intervention of the Agents to end the association between young Nazim Humayunjah and his grand-uncle Raushan-ud-daula, *see* AGG Melville to Pers. Sec., 3 August 1825, and reply, BPC, 19 August 1825, nos. 31–2.

[123] Nawab Nazim, Syed Mubarak Ali Humayunjah to Maj. Cobbe, 9 May 1833, F/4/1522/60090A.

[124] AGG Cobbe to Nawab Nazim, June 20 1833, ibid.

reported that Mubarak-ud-daula spent

> a great portion of his time... in the harum [sic], where he daily resorted in order to get rid of the importunity of supplicants or complainants who surrounded him in the morning. All access to him was denied, except thro' the intervention of the Eunuchs....[125]

In effect this resentment of the haram hinged fairly centrally on the inability of the officials to control the mothers and grandmothers who held sway until the 1820s. Reproductive success which could earlier lead to improvement in status for even the juniormost, declined from the installation of Amīrunnisa, when both eldership and motherhood were emptied of political value. The general devaluation of motherhood, in turn affected women in the lower echelons of the hierarchy.

By 1824, one British official rued the fact that the Government of Bengal was 'already sufficiently embarrassed by the *existence* of any such dignity and situation as that of Ja Nusheen of the Deorees *distinct* from and in a *great degree* independent of the office of Nawab Nazim'.[126] The slow erosion of the power of the gaddinashin was hastened by Amīrun's own childlessness, and the conflict between her and successive Nazims Walajah and his son, Humayunjah. In addition, agents of the Company tried to ensure that all potential Nazims would be 'educated' in a manner fitted to an English aristocracy, through the provision of English tutors for the younger males, and also through the Nizamat College in Murshidabad. Though Walajah, the father, had desired that his twelve-year-old son be educated in Arabia.[127] Government demurred due to the expense of the undertaking. From Humayunjah's accession, most Nazims depended upon the Company's agents for upholding their dignity and power. Encouraged to exchange one tutelage for another, young Nazims behaved in conflicting fashion, turning their backs upon their mothers, other matriarchs and the entire structure of the haram. By 1836, the conflict had turned inwards and for the first time, a Nazim turned upon his own mother, the head of his de'orhi, Najībunnisā, when she refused

[125] Home Misc. 584, p. 163.

[126] Letter to Mr. Loch, 8 October 1824, cited in Caulfield to Prinsep, 2 July 1838, *MNLI*, II, p. 214.

[127] AGG Ricketts to Pers. Sec., 3 January 1823, BPC, 5 January 1823, nos 73–4.

to acknowledge the two children whom the Nazim Humayunjah claimed to have 'reared' and one of whom he intended to nominate as his successor. Najībunnisā, following a pattern of complaint carrying deep symbolic resonances, accused the Nazim, her own son, of having sexual relations with a slave woman who had been both a concubine of his father, as well as his own wet-nurse.[128] The Nazim, in her view, had infringed both the moral and the political order: abusing the bonds of milk, and dishonouring his father by dishonouring his concubine. Elder males, like Raushan-ud-daula—a son of the Nazim Mubarak ud daula I and an uncle of two successive nazims—appeared to have reinforced the opposition of the mother to the succession of the boy chosen by Humayunjah. The Company's investigation into this was completed by September 1836, and it recognized the intended heir as 'legitimate' but followed it up with the condition that he be educated outside the zanana.[129]

From 1838 therefore when the nine-year-old Farīdūn Jah ascended the masnad, a military officer was appointed as the Superintendent of the Nazim's Education. Captain G.D. Showers, the first to hold this office, did not have a clear brief, and appears to have clashed constantly with the AGG, General Raper. Yet both men were completely agreed on the need to detach the Nazim from his mother Ra'īsunnisā (formerly Sahib Jan, the slave-girl of Moti Taifadar), and to this end decided to take the Nawab Nazim to Calcutta to meet the Governor-General, Auckland, without his mother. The fact that she had enjoyed both status and political control and the deference of her son, is attested by the testimony of the English officer who helped to conclusively break it down. With regard to young Nazim, Farīdūn Jah, he wrote

> His ideas were confined to the necessity of going through certain religious ceremonies and of paying proper respect to his mother and grandmother.[130]

[128] Translated letter from Najībunnisa Begam of 29 Rabīus San 1247 Hijri in BC F/4/1456/57367. For the symbolic significance of such a charge *see* F. W. Buckler, 'The Human Khilat', *The Near East and India*, 34, 903, 6 September 1928, pp. 411–22. Buckler had urged that the possession of the king's last wife or concubine constituted the capture of succession to the kingship, and 'going unto one's father's concubine' the prelude to taking over the kingdom.

[129] AGG Melville to W.H. Macnaghten, 23 September 1836, BPC, 4 October 1836, no.2, and 18 April 1838, no. 1.

[130] AGG Raper to G.D. Showers, 8 December 1843, *MNLR*, II, p. 461.

By 1841, Showers had compiled a list of days on which the twelve-year-old Nazim would be allowed to visit his mother and stay with her at night, and a separate list for those days when he would be allowed to go to the Killah but would have to return to the Mubarak Manzil.[131] This separation was not carried into effect 'without some opposition' on the part of the ruling matriarch, Ra'īsunnisā. Raper complacently expected that 'the angry feelings which she now entertains will yield to a just sense of the propriety' of such arrangements.[132] By 1843, the Nazim's sleeping in his mother's de'orhi was discontinued.[133]

The attempt to displace mothers meant also a concomitant attempt to distance each Nazim from the eunuchs of his mother's de'orhi. On a visit to Calcutta, the AGG prohibited the eunuchs of Ra'īsunnisā's de'orhi from 'attending the Nazim on occasions of public ceremony at Government House as... the presence of such persons might be regarded as an impropriety in European society'.[134] Upon finding them entrenched both in the affection and the household of the young Nazim, Showers made a virtue out of necessity. He decided to harness the power of the eunuchs 'in controlling the Nazim's habits'. The attempt to abolish completely the influence of mothers proved impracticable but it was tamed. In attempting this, the Agents themselves began to take over the political role of the mothers. In 1843, the AGG, Raper, tried to initiate a marital alliance between the Nazim and a daughter of the King of Awadh (and was snubbed by the latter).[135] The Company had indeed come a long way from the time of Cornwallis when, despite being pressed by a Mughal queen to negotiate a marriage with the household of the Nazim Mubarak-ud-daula, it had tactfully refrained from doing so. While the officers of the Company paid lip-service to certain ideals of kinship, they overrode the political significance of eldership and seniority where males were concerned, and motherhood where females were concerned. While each Agent, in turn, decried the conflict amongst kin, and resented

[131] Raper to Governor-General, 31 August 1841, BC F/4/2091/97365.

[132] Raper to Maddock, 20 October 1841, *MNLI*, II, p. 332.

[133] Raper to Sec. to Govt., BC F/4/2133/101356.

[134] Notes of AGG, 8 December 1843, BC F/4/2091/97365.

[135] Extract Bengal Foreign Progs. 23 August 1843, BC F/4/2091/97365.

the proximity of some slaves to their masters and mistresses, many of their policies encouraged the ruptures and fault lines within this group of kin and slaves. The transformation was completed when each in turn tried his hand at 'social engineering', through the manipulation and subsidies to specific kinds of marriages in the Nizamat. This was the most significant aspect of the departure that Raper had attempted—the transformation of the marriage-strategies of the Nizamat. To these we must turn next.

3

Making and Becoming Kin

From 1765, when the servants of the East India Company sought to legitimate their choice of allies in the 'revolution' in Bengal, issues of lineage politics were important. Vansittart, explaining his choice of Mīr Qasim, the son-in-law of the deposed Mīr Ja'far, as the Nazim, wrote that the death of Miran removed the heir-apparent of the government. Mīr Ja'far, the Governor wrote, 'had two sons by concubines, and a grandson, the child of his deceased son, by a concubine also;... they were incapable of taking care of the business, supposing the objection of their illegitimacy to be of no weight'.[1] From the outset of the relations between the Nazims and the English, other colonial contexts conditioned the constructions of genealogies in Bengal. In the reckoning of a slave-keeping English plantocracy, a non-slave could not marry a female slave because she was of a 'lower class' (and ethnically different). Slave-slave marriages, excluded from Church were characterized as informal or 'irregular'.[2] In both kinds of cohabitational forms, the female slave was denied canonical ritual and her progeny was 'illegitimate'. These attitudes begin to influence the consideration of lineage-claims in Murshidabad as well. Children born from slave-concubines without sacral rites were regarded as 'natural' or 'illegitimate'.[3] The test applied by the English was the

[1] Henry Vansittart, *A Narrative of the Transactions in Bengal from the Year 1760 to the Year 1764, during the Government of Mr. Henry Vansittart* (London 1766), I, p. 40.

[2] James Walvin, *Black Ivory: A History of British Slavery* (Washington, 1994), pp. 210–27.

[3] *See* AGG Cobbe to Sec. to GOB, Macnaghten, 24 June 1835, BPC, 27 June 1835, no. 12, for an application of the term 'illegitimate' to children who were 'the offspring of *khowassen* or concubines' not taken under any form of marriage.

presence or absence of a marital compact or ritual within these households. One of the charges levelled by Burke—against Hastings' appointment of Manni Begam to the Regency of Murshidabad–revolved around the absence of a marriage compact for her, 'sold as a slave; her profession a dancer; her occupation a prostitute'. In Burke's sense, her son Najm-ud-daula was 'this bastard slip', while the son born of Babbu Begam was 'the legitimate son'.[4] In fact, Burke was inaccurate: claims of marriage existed even for Manni Begam. In the certificate issued to Clive in 1767 for the five lakhs of rupees in money and effects, Manni Begam reported that the deceased Mīr Ja'far had instructed her to distribute the remainder of his effects 'after your marriage settlement is paid'.[5] Another petition of 1770 says that '30 Years prior to this, Meer Mohamed Jaffier Cawn... enter'd with me into the bonds of Matrimony, & entrusted Me with the whole Authority over the Mehal Serai...'.[6] However, from the late eighteenth century English officials in India were increasingly guided by Burke and Vansittart's gauges of relationships in the Nizamat. As a result a profound and fundamental elision occurred in the English representations of marriage and affinal structures. Just like Burke had supposed 'the lowest degree of infamy in occupation and situation' in characterizing Manni Begam's slave-status, other officials in India described female slaves in such ruling households as 'women of low extraction'.[7] Their reinterpretations of both Islamic and non-Islamic households alike were coloured by a deep-rooted contempt for the 'miscegenation' of slave and non-slave. English reinterpretations of cohabitational unions of men and women from different jātis as forms of such miscegenation rendered refinements of jural status within same-jāti unions redundant (see Introduction).

[4] E.A. Bond (ed.) *Speeches of the Managers and Counsel in the Trial of Warren Hastings* (London, 1860), II, pp. 22, 31–2.

[5] Translated Document Under the Seal of Meny [sic] Begum, 20 January 1767, p. 62.

[6] Progs. of the Controlling Council of Revenue at Murshidabad, 3–31 December 1770, in Walter K. Firminger (ed.), *The Letter Copy Books of the Resident at the Durbar at Murshidabad, 1769-1770* (Calcutta, 1919), II, pp. 95–6.

[7] For a description of the African and other slave *khowas* of the Mysore princes as 'the lowest description being Caffrees or low cast women from the Dekan', *see* Suptdt. Mysore Princes to Sec. GOB, 13 March 1837, BPC, 14 March 1837, no. 10.

The languages of 'legitimacy' suggested that the originating point of all progeny was some ritual of 'marriage', which then had to be found and ranked within a hierarchy of marriage forms, the highest of which were 'regular' and the lowest 'irregular'. Implicit assumptions that a woman of 'low' status was married in a similarly low-ranking, informal ritual were widely shared. In Kuch Bihar in 1865, when the Nazir Deo, Pertabnarain Kooer died, leaving older half-brothers, all of whom were described as 'my father's son by a slave-girl',[8] the Commissioner, unable to decide on the slave-origins of each mother, settled for the absence of a specific ritual to determine that,

> it is certain that she was not regularly married to his father, and that therefore her progeny cannot be considered as of equal rank with Ram Narain Kooer, who is the surviving son of Heminder Narain by his wife....[9]

What were these irregular or 'low' ritual forms? As illustrated from the judicial reinterpretation of marriage-terms like *phoolbibāhi*, and *gandharva*, in use in many native states in eastern India, flattening alone permitted their assimilation into 'Hindu law'. In 1814, the representative of the Dhenkanal state of Orissa had defined the phoolbibāhi as a 'Ranee of left-handed marriage' where another higher-ranked ritual form of marriage was practised for women of the same rank as the Raja.[10] However, the representative of Keonjhur had portrayed the taking of a phoolbibāhi wife as a last resort; when a Raja could not get daughters of the

> families of the Raja of Bishnupur, Kasurghar, Singhbum, he marries a daughter of the Bughel zat. Should the Rajah receive as a wife the daughter of a respectable person, not of his own caste, she is called a phoolbibahee.[11]

These answers suggested that the language of hypergyny itself was fraught with tension and could have two different implications: one

[8] Trans. Will of Protab Narain Kooer, Nazir Deo to Ramnarain Kooer, encl. in Dy. Commr. of Kuch Bihar to Lt.Col. Haughton, 2 June 1865, BPP, October 1865, no. 42.

[9] H. Beveridge, Dy. Commr. Kuch Bihar to Offg. Commr., 28 July 1865, BPP, October 1865, no. 47.

[10] *Twenty Five Questions Addressed to the Rajahs and Chiefs of the Regulation and Tributary Mehals by the Superintendent in 1814* (3e Calcutta, 1905), Answer to Q. II, p. 4.

[11] Ibid., Answer to Question VIII, p. 15.

that of a less elaborate ritual, and second, that of personal status characterized by respectable but lower jati-rank. By mid-century English officials and judges of the High Court had erased the dual meanings and pulls within the term. While in the case of one of the disputes in Keonjhur, the English counsel initially urged on the basis of his own long experience of the office of the Superintendent that a phoolbibahi was a concubine,[12] for the judges she emerged as the 'wife of lower caste'.[13] Similarly in Kuch Bihar in 1862, a woman described by dowager-matriarchs as a '*konyapatri* attached to the Raja's establishment'[14] and by the English officer initially as 'a concubine'[15] appeared in the will of the English-educated Raja as a 'wife married according to the Gundhorbo custom.'[16] Such strategies both normalized the relationship by assimilating it with Sanskritic models of high caste male and low-caste female (*anuloma*) marriage, and simultaneously erased distinctions between caste, race and slave-status. These were based on presumptions that 'lower' caste, jural status and 'menial' service were interchangeable: hence, finding a lower-ranked ritual form inferred the jural rank of the woman, and the latter did not have to be separately determined.

In translations and compilations of Islamic law authored by the Company's jurists, an identical representation, and ranking, of 'forms' occurred, conditioned by implicit assumptions regarding hypergyny and legitimacy. By the mid-nineteenth century, for instance, N.B.E. Baillie's translation of the *Fatwa-i-Alamgiri* tried to slot slave-wives within the ranked compacts allowed by different schools of Islamic doctrinal law. Thus he wrote

[12] F. Gouldsbury to Sec. GOB, 9 September 1861, WBSA, GPP, June 1862. no. 29.

[13] Ranee Bistooprea Patmohadea vs Basoodeb Dull Bewartee Patnaik, Guardian of Raja Dhunonjoy Bhunj and others, *Sutherland Weekly Reporter*, II, pp. 232–5.

[14] A.C. Campbell, PA to Commr. Assam to J.C. Haughton, 26 August 1863, WBSA, GPP, February 1864, encl., no. 51.

[15] J.C. Haughton, Offg. AGG, North-East Frontier to Sec. GOB, 21 August 1863, ibid., no. 49.

[16] Maharaja of Kuch Bihar to Maj. W. Agnew, Offg. AGG, 21 Srabun, 358 KB Sun, ibid., no. 42. In subsequent records, she is named as Nishimoyee Ayi, cf. Trans. Maharani of Kuch Bihar to Col. J.C. Haughton, n.d. BRP, June 1867, no. 101.

according to the Hanifites, the contract must be for the life of the parties, or the woman be the slave of the man; and it is only to a relation founded on a contract for life that they give the name of *nikah* or marriage. According to the Sheeas, the contract may be temporary, or for life, and it is not necessary that the slave should be the actual property of the man; for it is sufficient if the usufruct of her person be temporarily surrendered to him by her owner. To a relation established in any of these ways they give the name of nikah or marriage; which is thus... of three kinds; permanent, temporary, and servile.[17]

Instead of slaves coexisting within 'permanent' and 'temporary' cohabitational matrices, in Baillie's reconstruction of textual Islamic law of marriage, slaves had come to occupy a third category—servile nikah. This was the distillation of English colonial presumptions throughout the first half of the nineteenth century and was implemented during various investigations into specific households. These investigations were conducted on directives lifted out of a theological framework that was only superficially 'Islamic'.

In the case of such households, English civil servants' presumption that concubinage and the loosest forms of cohabitation were the *closest* relationship a female slave could have with a free master, completely overturned the law. It was the very antithesis both of Quranic prescription and of doctrines of 'legitimacy' adhered to by many local maulvis and qazis in Bengal in the early nineteenth century. In its prohibitions on the degrees of kin and affines who could not be married, the Quran specifically excepted slave-women (Sura IV.24) and called upon poor men to wed 'believing girls/from among those whom your right hands possess' (Sura IV.25 and Sura XXIV.32).

In Bengal at the end of the eighteenth century, Islamic jurists advising Company-controlled courts tried to reconcile Quranic injunctions with historical concerns and practice. For instance, in a dispute in 1798 in which a claimant of the half-share of the Serial zamindari of Mymensingh had based his claim upon a 'nikah' marriage between the zamindar Jafur Ali and his four slave-women, the judges asked the law officers to state the requisite forms of nikah. The maulvis began with declaring that no ceremony was necessary or binding since

[17] Neil B.E. Baillie, *Digest of Moohummudan Law, Containing the Doctrines of the Imameea Code of Jurisprudence on the Most Important of the Same Subjects*, (London, 1869), II, pp. xiv–xv.

'lawful enjoyment such as is obtained in marriage, accrues equally from the embrace of a slave girl'.[18] Textual law, 'legal acceptation', had limited slave-status to those 'taken from foreign infidels' but *practice* had resulted in 'slave girls, bought in times of scarcity, at a low price, from Moohummudans or infidel subjects', with whom therefore, modern lawyers, 'on prudential grounds', held marriage to be 'advisable' and 'preferable'. After setting down the requirements of reciprocal *ījāb-u-qubūl* (declaration and consent) the witnessing by two free, adult and sane men, or one man and two women with the same qualifications, they had even declared that the 'proofs' of marriage with the slave-girls—'the preparing of *sherbet* and *betle* [sic] leaf and giving the same to bridegroom and bride, and distributing a portion to the persons present'—were 'not requisite in law'. Even if the slave girls were not 'legally slaves but only commonly so reputed', the marriage with a fifth wife (a free woman), though invalid, would still require the payment of dower because it had been consummated; the child, even of this invalid marriage, would have his genealogy (*nasab*) established (and could not be bastardized). Shia and Sunni jurists alike frowned upon overhasty decreeing of illegitimacy: though it could occur as the result of *zinā* or fornication, the latter itself applied to a sexual relationship *beyond* that of the slave and master, and required very stringent evidence.

The maulvis' validation of marriage (not concubinage) with slave-girls appears to have been an attempt to reconcile practice in the eighteenth century with the classical textual requirements and meanings of nikah. Neither from an invalid marriage, nor from a slave-mother, were children to be considered 'illegitimate'. The attempt at weaving contemporary practices into the fabric of a 'traditional' law, characteristic of the maulvis above, was equally true of Hindu ritual specialists and local officers responsible for the *Pachees Sawal* (Twenty Five Questions). Both had identical ways of 'managing' historical practice by situating and incorporating it within a grid of formal strictures and injunctions. Hence, replying to a question regarding succession the officers of all the tributary states of Orissa

[18] *See* Gholam Husun Ali *vs* Zeinub Beebee (on the part of her son, Himmut Ali) in *Report of Cases Determined in the Court of Sudder Dewanny Adawlut*, 1791–1811, (W. H. Macnaghten, (ed.) Calcutta, 1827), I, pp. 48–52.

held that 'The son of a concubine, or slave-girl, has no right to the succession'.[19] However, later in the document, the opinion-givers accepted that *historically* slave-born sons had inherited kingdoms :

In the time of the Marathas it depended on the expenditure of money, and the side taken by the Sirdars and Paiks: hence it is that in every Gurjat the sons of slave girls are now on the guddee.[20]

Prior to the Company's intervention, wherever there were slave-born children, their belonging to, and often in, the lineage was never in question—disputes in such households revolved around juniority and seniority in claims to honours, privileges and exemptions, rather than legitimacy and illegitimacy. Hence superficially pro-slave opinions of English officials were based on a misrepresentation of law and practice within such households. An English official sympathetic to the claims of the slave-born sons of the Attia (Mymensingh) zamindar reported that though the mother of the claimants was a 'kunneez [sic] or slave or female servant... this is a difference of no consequence to the cause as by the Muhammadan Law legitimate and illegitimate sons are precisely on the same footing in point of succession'.[21] In terms of Islamic law, this portrayal of genealogies was simplistic because the category of 'illegitimate' did not refer to the slave-born at all. It also compressed multiple and discrete issues into one—the issues of jural status inheritance (slave/non-slave), the filiation of the slave-born (unilateral/ bilateral, matrilineal/ patrilineal) and the process of incorporation of female slaves into households and networks of kinship.

The attempt to distinguish among the children and women according to differing 'legal' compact then appeared to have formed the heart of a contradiction in colonial policies. If officials believed that slave-born and freeborn were all equally 'legitimate' these investigations were pointless. Nevertheless, they occurred consistently through the nineteenth century, and then again in the colonial courts in the second half of the century. The remaking of Islamic law and its texts at the hands of the Company's officers in the late eighteenth and

[19] *Twenty Five Questions* Answer to Question X, pp. 19–20.

[20] Ibid., Answer to Question XII, pp. 23–5.

[21] J. Duncan, Preparer of Reports at the Khalsa, to Revenue Dept., 24 February 1783, BRC, 25 February 1783, no. 28.

nineteenth centuries occurred on two different trajectories–one of legal understanding, and another of political and fiscal exigencies.

Contemporary Meanings versus Classical Prescriptions

The fundamental issues were the intersection of vernacular terms and local contracts with classical textual prescriptions and the ability of both English officers and the members of the Nizamat to manoeuvre the nodal points of this intersection. Clearly, much hinges on how marriage is conceptualized in a specific historical conjuncture: in contexts of polygyny, if female slaves could be, or were enjoined to be, taken as wives, were they differentiated from isogamous or homogamous (of like status, wealth, faith) wives by a differential scale of ritual, or the absence thereof? The problem in assessing the situation in Bengal was the wide difference in the way the same word—nikah—represented different meanings in the vernacular and the classical texts. Genealogies submitted in a dispute in the eight-annas share in the Attia zamindari in Mymensingh in 1778 describe a male ancestor of the house as having four wives, two by nikah and two by *sheady* (*shādi*).[22] Another man in the next generation in the same family, Khuda Nawaz Khan, was described as having a relation with a slave-woman of 'that kind of marriage called neka... but not that of Shady'.[23]

Contemporaries reported that the 'lower classes' believed that nikah was a lesser union and the ceremony of shādi enjoyed higher status.[24] G.A. Herklots, in his translation of *Qanoon-i-Islam*, noticed that,

> *Neekah* and *Shadee* are often used synonymously though in Bengal the former is only applied to a second kind of marriage called half-marriage. By the ignorant it is esteemed unlawful and disreputable, equivalent to keeping a mistress. Whereas, in reality it is the foundation of matrimony....[25]

[22] Enclosed in Council of Revenue at Dacca to W. Hastings and Council at Fort William, BRC, 3 February 1778, no. 110A.

[23] Deposition of Ramkunt, 15 May 1781, in F. Gladwin, Collr. of Silberris to J. Duncan, 22 June 1781, BRC, 25 February 1783, no. 28.

[24] Neil B.E. Baillie, *Digest of Moohummudan Law, on the Subjects to which it is usually applied by British Courts of Justice in India*, (London, 1865), I, p. 1, fn 1; *also see* J.N. Dasgupta, *Bengal in the Sixteenth Century* (reprint Delhi, 1989) pp. 92–3.

[25] Jaffur Shurreef, *Qanoon-i-Islam Or the Customs of the Moosulmans of India* (trans. G.A. Herklots, London, 1832), fn., p. 128; a heavily revised and re-arranged edition, by William Crooke, published in 1921, omits this sentence altogether.

Numerous cases in the courts of the East India Company in the 1840s and 50s bear testimony to the divergent status of nikah and *beah*, or shādī. For example, we hear from Dinajpore of a man whose wife by beah remained in her parent's house, while his wife by nikah lived with him.[26] Much more explicit was the charge laid by a woman from Purneah against a man who 'proposed to take her daughter in marriage by nikah but that she had as often refused compliance with his wishes, stipulating that if he wished to make her daughter his wife he must marry her in the proper and regular manner'.[27] The same distinction is sought to be warned against in one of the tracts issuing from the nascent movement to classicize Islam in Bengal in mid-century, in which Mullah Samiraddin says,

Shono re Allāhr bandā nekār khabar
Shādi nāhi kahe ketāb bhetor...
Nekār bayān Allah apani koyechhe.[28]
(O believers, hear the description of nikah
The Book says nothing of shādi
Allah himself described nikah).

Vernacular distinctions between nikah and shadi said little about the distinctions of textual law—nikah and mut'a—and also suggested that the meanings of terms like nikah were in flux throughout the first half of the nineteenth century. This flux came from tensions within the *practice* of slave-holders, rather than from the *prescriptions* of texts. The differences between contemporary, contingent, vernacular meanings of nikah and its classical textual attributes and consequences appears in turn to have confused English officials who investigated the *practice* within particular households through the prisms of Atlantic plantocratic and Islamic textual laws. To confound matters, indigenous litigants adapted to the language of the new rulers quite readily. While various local maulvis argued that the ritual marriage of a slave with a

[26] Government *vs* Gookoora Nusho, in *Report of Cases Determined in the Court of Nizamat Adalat*, (henceforth *RNA*), 1853, III, pt I, (Calcutta, 1854), p. 393.

[27] Musst. Guleeah *vs* Syed Babar Ali and Musst. Najoo in *RNA*, 1852, II, pt I, (Calcutta, 1853), p. 544.

[28] Maulvi Samiruddin, *Bedar Ghafelin* (Calcutta, 1877), p. 88. For a historical study of the movement, *see* Rafiuddin Ahmed, *The Bengal Muslims 1871–1906 : A Quest for Identity* (Delhi, 1988).

free master, or mistresses' son, was not a necessary precondition for civil rights of the slave and slave-born, there were indigenous non-jurists who contended that it was. For instance, in the course of the investigation into the antecedents of a potential Nazim at Murshidabad, one of the elder males of the extended kin-group had declared that ceremonies of feeding and prestations at pregnancy were 'not required by Law. By Law... marriage by "Muttah" or Nekah must take place....[29] These opinions were very close to, if not echoes of, various civil servants of the Company who had been using the term 'illegitimacy' to describe the status of slave-born children in the households of Bengal. Even those indigenous patriarchs who objected to the new dispensation did so with reference to the English authors. When the 'illegitimacy' of Badruddīn, of the Rassapagla-based branch of Tipu Sultan's household, had been communicated to its spokesperson, the latter replied

> Even allowing that as the English term it, Shah Allum alias Budderoodeen was illegitimate, yet by the Mahomedan Law all children are treated alike whether born (to use an English phrase again) in wedlock or not...
>
> As to the rights of children whether born by Shadee wife or others, see Sir W. H. Macnaghten's excellent work, Principles and Precedents of Mahomedan Law, page 85, Case 4—'The children by slave girls inherit equally with other children'.[30]

The unintended irony of this invocation of the English scholar-official should not be lost on the historian. For in 1836, Macnaghten the scholar had corrected the Superintendent for the latter's attribution of 'illegitimacy' to the 67 children born of the 213 khawās in the household. The term 'illegitimate', he said,

> used in consonance with the practice that has been hitherto observed in speaking of children not the offspring of the superior class of wives... would not seem to be appropriate *as children* by female slaves are equally legitimate

[29] Evidence of Shamsher Jang, 17 May 1836, IPP, 1 August 1836, no. 84.

[30] Prince Gholam Mahomed to Supdt. of Mysore Princes, n.d., IP&F, 26 October 1855, no. 87. W.H. Macnaghten, the son of an English lawyer who had served in the Supreme Court in Calcutta, derived his 'principles' from the cases, the 'precedents', that were submitted to the Qazi and muftis in the Sadr Diwani Adalat between 1814 and 1824, during which term he was the Register of the Court.

with the offspring of wives and the Law generally presumes the marriage of the parties where a man claim the parentage of a child born of a free woman....[31]

Yet as Secretary for the newly constituted Government of India Macnaghten had directed the enquiry in the case of the child claimed by the Nazim in 1836, asking for the existence of

any proof that the Mother of the Child was united to him by any of the forms of marriage in use among that sect of Mahommedans to which the Nawab Nazim belongs, or was the child born under such circumstances as to warrant his being claimed by the Nuwab Nazim as his offspring.[32]

Since in texts of Islamic law, there were at least two different 'forms of marriage', the question then splintered further. One usufructuary compact had a specific nomenclature; it was known colloquially as mut'a and was characterized by the fixing of a period or the fixing of a sum prior to the establishing of the relationship, closely resembling hire-lease arrangements. The recognition of this compact in Shia doctrine did not endow the children born of it, or the woman thus taken, with the rights of inheritance[33]. Sunni doctrine, on the other hand, while holding to the invalidity of this form, decreed that the genealogy (nasab) of the child of this union was established in the man, and by implication, to the category of heirs. Gholam Mahomed's reference to Macnaghten, the scholar-official, and not to arguments of qazis and maulvis given earlier in the century, also revealed how quickly indigenous litigants could adapt themselves to the requirements of colonial law, providing English-authored law texts to rebut English-authored determinations of household practice. Simultaneously, the divergences *within* English bureaucratic opinions were obvious; one set of officials linked 'illegitimacy' with birth from 'wives not of the superior class' (inchoately adhering to notions of hypergyny) and

[31] Sec. to GOB to Suptdt. of Mysore Princes, 19 July 1836, BPC, 19 July 1836, no. 20. Emphasis in original.

[32] Macnaghten to Commr. for Murshidabad Division, 5 March 1835, IPP, no. 173. For an illuminating discussion by the Commission of Enquiry (Oldfield, Macleod, and Melville) of the inapplicability of 'Islamic law' as constituted by Macnaghten's *Principles of Moohummudan Law*, see Report of 14 July 1836, IPP, 1 August 1836, no. 81.

[33] Baillie, *Digest*, II, p. 345.

others, like Macnaghten, who recognizing the inapplicability of such terms, nevertheless proceeded to convey and implement the decision of Government to debar the 'illegitimate' and their mothers (the slave-wives and concubines) from sharing in the Fund established for the maintenance of the household.[34]

For both Shias and Sunnis in Bengal in the first half of the nineteenth century, the relation between the slave woman and a man, either her master or one chosen by her master or mistress, could be comprehended within the term nikah, without necessarily exhausting the other cohabitational forms in which slave-women could be incorporated into a lineage. Yet all wives were not slaves, just as all female slaves did not become wives/concubines of their masters, or the males of the households in which they lived. The same stratification that existed within professions and skills within slave-based households characterized the stratification among 'wives'.

Marriage, Political Ritual and the Failure of Canon

While processes of kinship-making in the Nizamat were to be weighed against textual prescriptions by those dispensing the funds of the Nizamat, a gap appears to have opened up in their adherence to, and interpretation of, the taxonomies used. Officers in the first half of the century were not always trained in Persian and Arabic. Their use of these terms thus carried compound traces. For instance, an official deputed by the Government to Murshidabad, having described a woman as a '*munkooa* widow' then noted in the margin for his political superior that

> a munkooa wife is a wife of inferior rank, with whom the Ties of Marriage have been regularly performed, but without expense, or the observance of those external forms, which under the denomination of Shadee are customary on a public marriage, and which imply equality of rank between the parties marrying.[35]

[34] Sec. to GOB to Supt., 19 July 1836, BPC, no. 20. Further orders that children 'born of *khowas*' would not be entitled to pensions were reiterated in Sec. GOI to Suptd. of Mysore Princes, 26 October 1855, IF&P, 23 November 1855, no. 89.

[35] AAGG Monckton to C. Lushington, Sec. to Govt., 23 September 1817, BPC, 10 October 1817, no. 36. Underlined in original. The meaning of mankuh is 'legitimate wedlock'.

Another official commenting on the women in the household of one of the elder males found it 'doubtful whether the connection that subsisted between the Nuwab and these two females ever received any formal sanction, but in this respect they are on a par with the majority of the Ladies attached to the Nizamat, and may therefore, agreeable to custom, be termed *munkoohas*'.[36]

If mankuh denoted an inferior wife in this region, was there a specific term for a slave-concubine? According to some agents they were to be called '*mumtooah by nikah*' and yet others hired for a specific term were to be called concubines 'by *nikah o'mutah*'.[37] This official then went on to clarify that '*mumtood* described to me as a temporary compact inferior to nika, and renewable at the pleasure of the male party.' Watching and monitoring the problems in which the officers were mired, the household informants in the course of stipend-claims, or a succession dispute, used this terminology in order to imply status perhaps because it clearly was the favoured and assured route to success in achieving their ends. However, this use of classical taxonomies was paradoxical. For while indigenous litigants may have used a classicized language to represent particular relationships as nikah, they simultaneously revealed nikah to be the mirror opposite of its textual moorings, giving shadi a place that no text-book had. In the latter, nikah was formal; in local usage it was both informal and inexpensive, with little documentation and rarely any witnesses. Hence, in 1826, at the proposed nuptials of a new Nazim, Humayunjah, with his second cousin, Melville, the AGG, found himself at a loss regarding the amount of expenditure expected on such an occasion, since there was 'no precedent existing for any public ceremonial of the nature now contemplated'.[38] The absence of a ceremonial was significant precisely because contemporaries noted that no man, no matter how poor, gave his daughter in marriage 'where the projected union is

[36] C. Thoresby to Macnaghten, 5 December 1833, BPC, 19 December 1833, no. 100, BC F/4/1475/57976.

[37] Suptd. of Nizt. Affairs to Pers. Sec., 7 July, 1809, BPC, 15 July 1809, nos 65–66. The correct Persian word is either *mumtadda* or *mamtud*, both meaning protracted, extended.

[38] Melville to Stirling, 22 December 1825, BPC, 6 January 1826, BC F/4/1040/28656.

with an equal, without a... parade of music, and a marriage-portion in goods and chattel'.[39]

Where the Company's agents looked for the marriage-rituals of canonical law, members of Islamic households depended not on 'legal compact' or its nomenclature but on other markers. Belonging in the lineage was fundamentally premised upon clear proof of *title* to that slave; in other words, whose slave the woman had been. As Tahauwur Jang, the younger sibling of a deceased grandson of Muhammad Reza Khan put it, what was critical to establish was that the slave who had borne children was *not* a slave belonging to the man, but one belonging to the mother of the deceased male, so that her child could be claimed as the household's slaves rather than that of its sahibzada.[40] The maulvis referred to by him were clear about this: asked about the 'legitimacy' of a child born of a slave girl 'belonging to another individual', and not to the father, opined

> The legitimacy of a child depends upon either of these two Circumstances.... The birth of it by a lawful wife of its Father and 2ndly by a slave girl of his own, a Fact which is as clear as the Sun and accords with all the Books of the Law of Inheritance.[41]

Similarly, those elder males who contested the 'sonship' of Faridunjah in 1835–6, did so on the basis not of the slave-origin of the child, but on the *failure* of the ruling Nazim to establish his 'possession' of the dancing girl who was the slave of Moti, of a *taifa* of skilled women and girls.[42] Clearly, it was not 'legitimacy' of a child that was in question, but the different bases for establishing claims over the women, and children born of these women. As the contestants in Murshidabad in the 1830s urged, there were very precise codes by which the possession by a Nazim could be signalled. Especially when

[39] Mrs. Meer Hassan Ali, *Observations*, p. 185.

[40] Enclosure no. 1, Nawab Tahauwur Jang to Sec. to Govt., recd. 22 August 1835, BPC, 10 September 1835, nos 30–1.

[41] Ibid.

[42] Nawab Raushan-ud-daula and Nawab Zulfikar Ali Khan to the Governor-General, recd. 20 Jan. 1835, IPP, 5 March 1835, no. 172; Raushan-ud-daula to Dy. Gov. of Bengal, n.d., NAI, FPP, 21 Nov. 1838, no. 56. This was apparently common knowledge, judging by articles in the contemporary press, vide *Moorshedabad News*, 13 October 1838, and *Daily News*, 9 October 1838.

the woman was a slave-girl of another, as many of the dancers who became concubines of previous Nazims were, it was said

> if any khawāss, nautch girl or other should find favour in the eyes of the Nazim, she being then designated khādima, and sent into the Mehal, has no liberty to go elsewhere... then an allowance is fixed for her food and clothing... [they] had no longer the power of going to their former mistresses.[43]

Explicating a code in which the proximity of a slave-woman to the Nazim was marked out by the degree to which she partook of the spatial and social seclusion that characterized his person, and partook of his beneficience, such petitions could not be reduced to fit the framework of 'legal compact' that officials tried to bind them within. There were specific non-canonical rituals that marked the 'wifehood' of women in the Nizamat. As one gaddinashin, Amīrunnisa Begam, explained her conferring titles and honours upon a woman

> It has been a long established practice in this Sirkar, to confer titles and Soojnees upon all those persons who become respectable by bearing children to the principal members of the family, or by ranking in the class of wives,... in order to make them equal with others'.[44]

This kind of political ritual that endowed a woman with 'wifehood' represented the exact inversion of the trajectory that determined that the legitimacy of a child followed from the status of a ritual. This was equally true for other ruling households. Motherhood distinguished the transition of a slave-companion to the status of wife, rather than followed upon it. In Tripura, where successorship had come to a head in the mid-nineteenth century, the Principal Sadr Amin as well as the Calcutta High Court employed the criteria of particular rituals of marriage to determine 'legitimacy' and disqualified one contender on the grounds that his mother was the 'Sebaita (handmaid) Kachooa... a mere servant who prepared betel, supplied tobacco, and performed other similar duties in the apartment of the second Ranee'.[45] However, the testimony of the female elders of the household suggested that it

[43] Abstract trans. of letter of Nawab Raushan-ud-daula, 10 Bysack 1243 B.S./ 22 April 1836, IPP, 1 August 1836, no. 89.

[44] Enclosed in J. Caulfeild, AGG, to the Nazim, 26 June 1838, *MNLI*, II, pp. 210–11.

[45] *Sutherland's Weekly Reporter*, I, Civil, pp. 194–5.

was not ritual but reproductive success that had endowed the women concerned with rank. As one of the elder females said of the preceding Raja, 'Eshan Chunder's mother was Kachooa before, but after the appointment of Eshan Chunder as Joobraj, coins having been struck in his name, she was made a Ranee'. Another likewise urged a similar elevation in status for other *kachooas*, in the statement 'After the birth of Chuckrodhuj Thakoor, she was called the Thakoor's mother'.[46] Judicial and other officers tried to distinguish between women who were married by a ranked ritual, and those who were not, within households. This inverted the moral order and socio-political hierarchies. Some women could be raised to the status of ranis as a result of their motherhood, just as certain mothers were to be elevated to the rank of wives in the Nizamat. A ritual marriage was *not* the key to changed statuses of women within a household, in as much as slave-born sons were preferred by many rulers and chiefs.[47] Furthermore, the language and grammar of this elevation was specific to each household—the use of a particular form of jewellery or wearing apparel, or form of transport, or the use of a title. As a scion of the Nizamat, Ali Kadr and heir-apparent to the masnad, put it

> The Nawab Nazims were in the habit of raising their favourite ladies to the dignity of wives by merely pronouncing them to be such, and directing that they may thenceforth be styled Nawabs by conferring upon them musnud seat and other accompaniment [sic] honours.[48]

These signs constituted a grammar of political ceremonial, equivalent to the granting of khil'ats. In the Nizamat, when a woman became a mother, she acquired a title. The changes in names and the accretion of titles thus marked the movement of women into rungs of intimacy, of proximity to various males and masters/mistresses, as well as in reproductive status. Like male slaves were named after seasons (Basant), legendary heroes (Dārāb) or gems (Zamurrud or emerald), the first rung of incorporation for female slaves too was marked by a name

[46] For lists of Rajas of the state of Hill Tipperah, *see* Memorial of Information Regarding Native Chiefs in Bengal, L/P&S/F/76/2, no. III, p. 8.

[47] L. Griffin *The Law of Inheritance to Chiefships as Observed by the Sikhs Previous to the Annexation of the Punjab* (Lahore, 1869), pp. 54–55.

[48] Ali Kadr to the AGG, 4 April 1877, BPP, January 1878, File 17 B, nos 28–9.

which either valorized a particular attribute of the girl (Gulbadan or flower-like, Hansmukh or of the smiling face) or sought to cultivate it in her by association with a jewel (Pokhraj or topaz), fragrance (Elaichi or cardamom) or flower (Sugandharaj). Some of these females would experience another name gain subsequently, depending upon their ability to please their owners or males of the household. Speaking of the customary naming practices of nobles and grandees, an observer had remarked that they

> bring up other people's daughters, have them taught dancing and singing, and such are called *gaeenen* (singers) to the end of each of whose names they add the word *Baee*... and when they make a favourite of one, they from affection honour her with the title of *Khanum* and if they are devotedly attached to her, dignify her further with the appellation of Begum. Their slaves in like manner they call first *Booboo*, and when they cohabit with such a one, honour her with the titles *Baees*, *Khanum* or *Begum*.[49]

The significance of name-changes and status-changes where slave-women were concerned was evinced yet again when a slave-concubine of Dilawar Jang (the son of Muhammad Reza Khan), stopped using a seal bearing the name 'Booboo Goolaub' and had a seal styling herself the 'Mother of Mohummud Cauzim'.[50]

Between 1800 and 1880 many slave-women came to bear different titles in the course of their lives depending not on, or the kind of, marriage-ceremony, but on their reproductive success. In the 1870s, in the course of an investigation into the status of certain sons and daughters of the Nazim revolving entirely upon the form of the ceremony of nikah, the Agent asked the mother of the reigning Nazim, Ra'isunnisa Begam, whether the position of the late Mihrlakha Begam was that of mankuh or *mamdūd*, to which she replied that it

> is the custom in the family on the occasion of a nika to send garlands, scents, and pan round to all. It was on these appearing, and my inquiring of the nazir (eunuch) of my deorhee, Firoz Ali Khan, why they had been sent, that he told me they had come in consequence of the nika of the nawab Nazim with Husseena... another reason why I gathered he had entered into nika with her and Mujleh Saheb's and Amir Saheb's mothers is, that it is only the

[49] Jaffur Shurreef *Qanoon-e-Islam*, p. 17.

[50] Pers. Sec. Govt., to Suptdt. of Nizt. Affairs, 5 October 1810, *MNLR*, II, p. 34.

custom to give the 'nath' [...] and 'baina' [...] to Munkooahs; and much later, when the Nawab Nazim brought all three to me to present their nuzzers, they all three were wearing those ornaments. This was before the *shadi* (marriage) with Nawab Shumsejehan Begum.[51]

The chronological order in which nikah and shādi were performed was distinct. The private acts of elevation to mankuh-dom occurred *before* the public and expensive ceremony of marriage with a cousin, Shams-i-jahān (the grand-daughter of Jigri Begam, a daughter of Mubarak-ud-daula I[52]). The significance of the chronological order was that nikah could be performed with a slave-woman in the keeping of a senior matriarch. Hence, one of the women, Hasīna, described by the mother of the Nazim as having been taken in nikah by her son before his shadi, was

an African slave in my personal service... I had made arrangements for her marriage... when I discovered that my son, the Nawab Nazim, who was still living in my quarters (deorhee), had entered into a liaison with her. I also found he had entered into a liaison with the mothers of Mujleh Saheb and Amir Saheb (second and third sons), who were also two of my personal attendants... the Nawab Nazim... carried all three off to his own quarters... afterwards he begged of me Haji Begum, who had been entrusted to my charge by her father... she is now the Malika Zumaneah Begum.[53]

As we can infer from this, the marriages of males of the Nizamat combined apparently isogamous (like with like) and expensive unions called shadi with quiet and inexpensive arrangements like nikah or mut'a. However, the slave-marriage came chronologically prior to the isogamous marriage. What was more crucial, though unstated in this specific instance, was the fact that Hasina gave birth to a son prior to the Nazim's isogamous marriage in 1846. Thus the slave-girl's entitlements to the symbols of status, like the title of Mihrlakha, the

[51] Cited in Report by W.J.W. Muir, 4 May 1877, in BPP, January 1878, File 17B, nos 28–9, Appendix B.

[52] AGG Raper to Sec. GOB, 31 July 1845, BC F/4/2185/106670.

[53] Cited in report of AGG, W.J.W. Muir, 4 May 1877, BPP, January 1878, nos 28–9. The African slave-origin of Hasina was well-known to English officials before this investigation, *see* AGG Thomson to Sec. GOB, 25 October 1864, NAI, FPP, December 1867, 124B and Offg. Poll. Agent, Turkish Arabia, to Offg. Sec. GOI, 7 April 1870, NAI, FPP, June 1870, A 126.

nose-ring, the *baina* (ornament for the forehead), the seals and the *sozni* (a special cloth for seating), followed from her bearing a son to the Nazim rather than any individual ceremony of marriage. And as the description by the mother of the Nazim revealed, intra-household conflicts could arise because of the breaches of the symbolic order: the Nazim had performed the nikah with three women in his mother's keeping *without* the active cooperation of his mother, and left her to infer from the signs on their bodies, that they had been elevated to the status of wives.

Were the elaborate, formal rituals of marriage then politically insignificant as far as ruling males, and his half-brothers or paternal uncles were, concerned? If we assess the relative absence of shādi and its affinal consequences from the lives of the sons and uncles of the Nazims, it appears so. The consanguinous males whom we have detailed information about are those males who were accepted as sons of Mubarak-ud-daula I. Of the thirteen, there were only two whose formal marriages (shādis) left any imprint on the records. Of these, Nasīruddin Haidar came nearest to establishing an affinal relationship with an important and powerful family, 'of rank and respectability' for a short while in 1811–14. After the death of his grandmother Babbu Begam, he went to Lucknow, where he was married to a daughter of Madar-ud-daula; hence he became *hamzulf* of the Nawab Vazir himself who was married to another daughter of Madar-ud-daula.[54] However, this important affinal tie proved to be very short-lived, and the groom extremely improvident. Nasīruddin Haidar quarrelled with his wife and left Lucknow within three years. The only other brother who came close to matching Nasīruddin Haidar's affinal network was Mīr Mehdi, another son of Mubarak I, who married the great grand-daughter of Mīr Ja'far, Gauharunnisā, at considerable expense.[55] It is possible that the rarity of elaborate and formal marriages, shādi, itself distinguished the successful and wealthy

[54] Pers. Sec. to Suptdt. of Nizt. Affairs, 19 February 1814, *MNLI*, II, pp. 128–9 and AGG to Pers. Sec. 2 November 1822, BPC, 14 December 1822, no. 22.

[55] Actg. AGG to Actg. Pers. Sec. 7 May 1818, Extract Poll. Cons. 22 May 1818, no. 92, BC F/4/618/15418; Mīr Mehdi to Governor General, recd. 24 December 1823, BPC, 27 February 1824, no. 46 cf. the contracting of a heavy debt at his marriage; AGG to Sec. GOB, 14 February 1826, BC F/4/1040/28658.

male from the one without the wealth to sustain affinal connections.

However, what united all the males in one generation was their equal access to slave-girls provided by their mothers or important matriarchs in the de'orhis. This role of the haram in providing sexual partners for young males was not new. Siraj-ud-daula's principal wife was the daughter of the courtier Iraj Khan, and that wedding was celebrated in great splendour[56]; yet it was Lutfunnisā', (a slave girl originally named Rajkunwar) reared and gifted to him by his mother, Amīna Begam[57] who was his 'companionate' wife. This pattern persisted in the lives of most of the sons of Mubarak I. Mīr Mughal was described as having 'four harams... presented... [him] by the Babbu Begam for his service'.[58] Yet another brother, Saiyid Ali Reza urf Mīr Maulvi, had two slaves provided by his mother, another given by another senior matriarch and he himself purchased another female for two hundred rupees.[59] As revealed by the lives of these sons of Mubarak-ud-daula I, the companions of their adolescence were also other females reared in the de'orhis. In adulthood, the males' access to higher-status or expensive marriages (with promises of *mahr* payments) increasingly depended upon the wealth they could command as well as the affective relations they could build with other men, including English Agents and Nazims alike. Both goodwill of the latter and wealth (stipends, inheritance and so on) were in very short supply where these men were concerned; most sons of a previous Nazim died in abject poverty. So while the sexual proximity of some female slaves to males in the Nizamat distinguished them from other women in the de'orhis, and motherhood separated the privileged

[56] Most histories of this period mention the grand celebrations occurring over several days of the weddings of Siraj-ud-daula and his brother Akram-ud-daula in 1746. For a description of this event, which acquired legendary status, *see* Charles Stewart, *The History of Bengal* (reprint Delhi, 1971), p. 473; Karam Ali, *Muzaffarnamah* and Yusuf Ali *Ahwal-i-Mahabat Jang* in J.N. Sarkar (ed.) *Bengal Nawabs* (Calcutta, 1952) pp. 36–7 and pp. 117–18 respectively.

[57] *Seir-ul-Mutakherin*, I, pp. 94, 187; Sirajul Islam (ed.), *History of Bangladesh 1704–1971* (Dhaka, 1992), III, pp. 41–2, Rafiqul Islam *Daccar Katha 1610–1910* (Dhaka, 1982), p. 101. Later historians considered her to be the ideal wife, cf. Bandopadhyay, *Banglar Begum*, p. 7.

[58] AGG to Actng. Pers. Sec., 23 September 1817, BPC, 10 October 1817, no. 36.

[59] AGG to Pers. Sec. 8 October 1817, BPC, 17 October 1817, no. 52.

women from the untitled ones, the simultaneous access to both isogamous and slave-wives distinguished successful senior males from juniors in the household.

It appears that the factors responsible for a lack of precedent for expensive marriage ceremonies in early nineteenth-century Nizamat were both financial embarrassment, as well as this particular aspect of the haram. What prevented a public celebration, a shādi, of a consort to the Nazim, or his brother, was the absence or attenuation of her own natal kin in the ceremony. This is where the quietness and cheapness of nikah, the social inferiority of slave-wives and the rule of the matriarchs coincided. Where marriage expressed both a reciprocal acceptance of the other party's claims and a bond of identity between them, with bride-receivers accepting the equal status and claims of the bride-givers, isogamous marriages entailed bilateral obligations and a certain amount of ceremonial deference to the equal-status wife and also a formal etiquette involving ceremonial and ritual prestations and honour towards the household from which she came.[60] A first generation slave-wife or concubine did not carry with her identical political aspirations that the family of a woman of the indigenous elite might have entertained, or the political leverage that it might attempt to exert. Indeed, precisely because of her lack of kin or the inability of her kin to claim any prerogatives of kinship, the Nazims would not have to defer to any other family, ceremonially or otherwise. Such kinlessness was so attractive that many non-Islamic hegemons in other parts of Bengal tried to ensure it in their marriage strategies. For example, in the ruling household in Tripura, P.C. Mahtab noted that 'new-born female children', made to 'lose all basic connections with their previous relatives and friends' were taken into the household and reared as *kaccha* brides. This simultaneously released the grooms from affinal obligations, and any children born of competing bilateral claims.[61]

In looking for the textual pre-requisites of marriage, British Agents thus considerably under-estimated the political role of the mother, or

[60] Michael H. Fisher, 'Political Marriage Alliances at the Shìi Court of Awadh', *Comparative Studies in Society and History*, 25, 1983, pp. 593–616.

[61] P.C. Mahtab, 'The Bengal Nobles: A Status Group, 1911–19', *BPP* 92, 1973, pp. 26–7.

senior matriarch, in every male's life, and the material stakes that conditioned these processes of the life-cycle. Evidence by males taken in the late nineteenth century tended to eliminate the fact that earlier in the century such acts of recognition, and elevation, had to be that of the matriarch in charge, rather than a male. The erosion of this political role of matriarchs particularly after the 1840s erased it from historians' visions as well. However, earlier in the century, matriarchs had played a very significant role in providing wives for the junior males in their de'orhis. For it was in order to generate the kinds of claims that indigenous elites might have exerted that each gaddinashin Begam of a de'orhi 'reared' slave-girls, who would then be 'given' to a particular male of the household. Therefore, each Nazim's, and other males', access to 'wives' till 1840s was predicated upon the processes of filiation and social reproduction within the haram—processes that to the English looked like 'adoption'. It is to that that we now turn.

Making Kin: Political Practice against Canonical Injunction

The dominance of the ideal of motherhood expressed two separate issues as though they were one: filiation of the child born of the slave, and claims of authority over both the slave-woman and her child, as over all other freshly acquired or first generation slaves. Despite the fastidious attention that individual British officers paid to issues of Islamic jurisprudence, few noticed the anomaly of describing various children within the mahals as 'adopted' daughters and sons. In the strictly textual tradition of Islamic laws, adoption is explicitly abrogated.[62] Yet between 1770 and 1850, the records of the household seem to abound with 'adopted' sons and daughters of various women.

What was the substance of these adoptions and were they all alike? Any consideration of such 'adoption' has to take into account the specific situation within the household at a particular moment and generation, as well as the broader question of inheritance of jural status where slaves were concerned. Sonship and daughterhood implied specific kinds of claims and obligations within the structure of the

[62] Mernissi, *The Veil and the Male Elite,* pp. 175–6.

household. So when Manni Begam was described by Valentia as having 'adopted' a son, he did not fail to mention that the timing of this adoption coincided with a peak in the conflict between the ruling Nazim and the Begam.[63] This adopted son appears in a record of 1818 under the name of Mīr Ghulam Ali Achay Sahib, Qaim Jang. Though married to a daughter of Ihtiram-ud-daula, (brother of Mīr Ja'far) his children, 'two sons' born of a concubine were 'brought up and educated' by Manni Begam, and their women in turn continued to serve the Begam and live in her de'orhi.

The antecedents of Qaim Jang provided ample illustration of the complex intergenerational movement of some slave-born children into the group of 'kin'. For his history was tied up to the daughters that the British officials had overlooked in 1765 when counting the 'family' of Mīr Ja'far. One such daughter, Misri Begam, born of a concubine other than Manni or Babbu, in turn had had a daughter, Jan Begam, whose marriage with Mīr Asad Ali had produced three sons, one of whom was 'adopted' by Manni Begam. Evidently, this selective adoption was a political 'manoeuvre of the Begum whilst she was at variance with the Nawab'.[64] Similarly, in mid-century Keonjhur, witnesses examined during a disputed succession declared that Dhunonjoy, born of a maid-servant and a *paik* of the Raja's household, had been abandoned by the natural parents and been reared in the Raja's household by his mother ('out of charity taken care of by the Raja's mother').[65] The Superintendent dismissed this as a 'monstrous improbability': yet it was a fairly characteristic pattern in Bengal. However, if particular adoptees could have posed a challenge to a succession or in Qaim Jang's case, to the Nazim, other such adoptees did not. Thus Najm-ud-Daula, Manni Begam's uterine son, was reputed to have 'brought up' a boy named Ramzan Ali, who was clearly listed as a chela.[66]

[63] George Mountmorris, Viscount Valentia, *Voyages and Travels in India, Ceylon, the Red Sea, Abyssinia and Egypt in the Years 1802–1806* (London, 1811), I, p. 184.

[64] Description by Monckton in Doc. K, *Matfarqat Sherista*, BPC, 6 July 1816, no. 51.

[65] Order of the Court of the Superintendent of the Tributary Mehals in Dhunonjoy Bhunj *vs* Brindabun Bhunj, 19 March 1862, encl. in Ranee Bistoprya Pat Mohadayee to Sec. GOB, 29 April 1862, WBSA, GPP, June 1862, no. 21.

[66] For Ramzan's name on the list of chelas, *see* Doc. M, BPC, 6 July, 1816, no. 51.

Apart from the political moment at which Manni Begam 'adopted' Qaim Jang, there were aspects of reproductive politics within the haram which explained an ambitious matriarch's dependence upon this mode of rearing an heir. As we tried to argue earlier, the slave-concubine's access to power, dignity and privilege rested not always upon a 'ritual' of marriage but upon her bearing of children to her master. Yet, becoming a birth-mother depended upon a host of interlocking factors within the de'orhis. Though records of actual birth and contraceptive practices are scarce in the official records, what is in evidence is the frequency of infant-mortality through the nineteenth century. Coupled with the travel writings of earlier centuries in the case of the Mughal households, this frequency of infantile deaths awaits investigation. Manucci, for instance, stated that it was 'usual for Mogul princes to rear only four of their sons', and when one of Shah Alam's (Bahadur Shah) wives, Nurunnissa, became pregnant, she 'took drugs to procure abortion. What she had taken produced no effect, and in her husband's presence she asked me for drugs for that purpose'.[67] Though Manucci believed that the child was eventually poisoned, it is not for that reason that his evidence is interesting. The way it is recounted indicates firstly, that knowledge of abortifacients may have been an established feature of the Mughal haram and secondly, the concentration on male infanticide. Both can be explained in the context of the politics of motherhood. Later historians have noted that Nurunnisa did have a son called Rafi-ul-qadr (subsequently called Rafi-us-Shan).[68] This either falsifies Manucci's claim or forces us to reinterpret his story in the light of a different perspective.

Data on mothers in the Mughal household is not systematically collated, yet whenever children and mothers are discussed together, especially for the eighteenth century, the inference is that some kind of delimitation to single sonship was followed. The episode recounted by Manucci may make sense only if we understand that single-sonship may have been particularly important to that household precisely because the role of a mother was an important political one. Various accounts touch upon the sagacity or diplomatic skills of the mother of a prince, especially at key moments of succession, or in battle.

[67] *Storia do Mogor*, II, p. 384.

[68] W. Irvine, *Later Mughals* (ed. J.N. Sarkar, Calcutta, 1921), p. 141.

Thus Muhammad Muazzam's mother Nawab Bai 'gave good counsel to the Prince, and strongly dissuaded him from yielding an assent; and from giving any aid, assistance or intercession on behalf of the Rajputs', while passing through the Rana of Udaipur's territory.[69] Similarly, Farrukhsiyar's mother is mentioned as having first negotiated with the mother of one of the Saiyid brothers for his support in the battle of succession. This was equally true of the subah of Awadh: Shuja-ud-daula's mother deployed considerable diplomatic skills and personal wealth to stem an incipient revolt among his officers in 1754. Shuja's son, Asaf-ud-daula, too succeeded his father only because of the intercession of his mother.[70] In polities, where the mother was so critical a political personage, no woman could be allowed to have claims upon her authority contested between two uterine sons. Hence, in Awadh, despite the fact that Shuja-ud-daula had 22 sons and 34 daughters, the mother of Asaf-ud-daula was reported to have been the mother of one son alone: she explicitly acknowledged this to her mother-in-law 'I have had but the one son in my whole life; bad or good, he is my sole treasure'.[71]

In the early years of Mīr Ja'far's household, a similar policy of single-sonship for each of the concubines is decipherable, though there was no policy of limiting daughters. We can begin with the concubines of Ja'far himself: to Manni Begam was born Najm-ud-daula, to Babbu Begam, Mubarak-ud-daula (and possibly one daughter). The same policy appears to have prevailed in the haram of Mubarak-ud-daula I: for the names of the concubines we do have, there are only the single sons listed against the names of each, like Sharīfunnisā Khanum, mother of Mīr Mangli, Lutfunnisā mother of Mīr Mehdi, and Mubarakunnisā Khanum, the mother of Mīr Muhammad Ali and Jigri Begam.[72] This pattern began to falter in the next generation: for example, despite access to slaves, concubines and others, many of the sons of Mubarak I died childless, or left behind 'adopted' sons and

[69] Khafi Khan, (ed. Eliott and Dowson), VII, p. 300.

[70] Barnett, *North India between Empires*, pp. 44–5, 100–1.

[71] Ibid., fn.9, pp. 100, 101.

[72] Both British officials and subsequent historians paid no attention to these finer details, and confused counts beginning with the former passed into the historiography as in the case of Najm-ud-daula and Saif-ud-daula, both of whom were described as sons of Munni Begam. Similar errors exist in official accounts of

daughters as their heirs. Mīr Mughal nominated his sister's daughter. Mīr Maulvi 'adopted' the concubine-born daughter of a predeceased half-brother called Saiyid Hussain.[73] However, wherever we have evidence for the other sons, there a pattern of single-sonship for the concubines also followed. For instance, Pyarunnisa, one of the two women an agent had described as mankūha and whom the Nazim described as haram of Ab'ul Hasan, was the mother of a three-year-old boy.[74] Well into the nineteenth century, such a pattern of single-sonship appears to have been practiced in the easternmost states of Bengal, like Kuch Bihar: in the household of the Raikats in Jalpaiguri, Surbodeb Raikat died in 1848 leaving Makaranddeb, son of slave-girl Jamonee, a younger son Rajendrodeb, son of Bissessuri and a third Fanindrodeb, son of Dhulpi/Rutnessuri.[75] The political implications of a delimited motherhood of sons was that though technically brothers, through a common father, each male was the lone son of a mother, whose ability to generate support for her son and for herself conditioned the dissensions among brothers, uncles and affinal males. Single-sonship was both an admission of the importance of the mother to structures of political authority, as well as a systemic curb upon her power.

Very little is known of how such delimitation occurred. Was it purely accidental, or were there actual strategies and practices, like foeticide, which were deployed to select and prune male infants of

---

the sons of Mubarak-ud-daula I and their mothers. The only corrective when two sons are listed against one mother is to check the life histories of the daughters of the same mother, from where only one emerges as 'own brother'. One instance of this is the two sons ascribed to Lutfunnisa, Mīr Mehdi and Abul Hasan, by AGG F.W. Russell in letter to Prinsep, 25 June 1821, BPC, 7 July 1821, no. 78. However, both Russell's own subsequent description of the family of the daughter, Rahīmunnisa, and earlier descriptions from the period 1809, refer only to Mīr Mehdi as her brother.

[73] Will of Sayyid Ali Reza in Suptdt. of Nizt Affairs to Pers. Sec., BPC, 20 August 1813, no. 61; AGG to Pers. Sec. 25 October 1813, BPC, 6 July 1816, no. 62. This daughter was described as such in Letter from Nawab Abul Qasim Khan (Mīr Mangli) and others to Mr. Rocke, n.d. in Suptdt. of Nizt. Affairs to Pers. Sec., 15 November 1810, BC F/4/372/9258.

[74] AAGG Thoresby to Sec., 5 December 1833, Extract Pol. Cons. 19 December 1833, no. 100, BC F/4/1475/57976.

[75] Later historians referred to Makaranddeb as 'Gopkanyar garbhajata' or born of a woman of the Gop jati, *see* Amanatulla Ahmed, *A History of Kuch Bihar (in Bengali)* (Kuch Bihar State, 1936), I, p. 231.

specific mothers? The suspicion must rest with the latter. Particularly since such sonlessness, and indeed a general childlessness, appeared to have characterized the same-status, or higher-status wives through a large belt of territory in India in the early nineteenth century. In some, distancing strategies between spouses were in evidence, as with the Raja of Birbhum, Muhammad Zaman Khan. In 1794, his 'royal' wife, the daughter of the Raja of Kanakpur, had come to stay with the Raja's mother; 'after continuing here for six months and the Raja never going near her',[76] she resided with her own father. Asked by the Court whether he had ever 'heard of the head Ranny after coming to her husband's house returning to her family during her husband's lifetime', one witness had replied that this indeed was the custom.[77] This impression is strengthened from the list of the Raja's household submitted after his death, which is headed by the 'married Ranny' who has no children listed against her name but one who is termed 'neckye' [nikahi] and eleven others are shown to have been mothers of children between the ages of one and eight years old.[78] The systemization of this pattern is again suggested by the fact that the previous Raja, Bahadur Zaman Khan, had himself been born to a slave-girl belonging to Asad-us Zaman Khan's mother.[79] Scattered fragments from autobiographies of *bhadralok* Bengalis later in the century suggest the role of female servants in the employ of the matriarchs, and the active controls of conjugality by the mothers of grooms in ensuring celibacy or reproductive delimitation.[80] If this was true for households of smaller means, fewer members, and lesser financial and material stakes in the reproductive success of particular women, it was much more feasible in ruling and hegemonic

[76] Evidence of Buddun Metre, in Extract Progs. of the SDA, 6 July 1796, J.H. Ernst, Actg. Collr. of Birbhum *vs* Raja Mahomed Zaman Khan, BC F/4/19/767.

[77] Evidence of Motyollah, ibid. One of the sons of Kalyan Singh, the Naib Nazim of Bihar, urged a similar pattern of homogamous wives living apart from husbands, and the latter residing mainly with slave-concubines, in letter to Govt., 1 April 1823, BPC, 9 May, 1823, no. 157.

[78] List submitted by W. Cowell to Board of Revenue, 13 April 1802, BC F/4/167/2939.

[79] John R.Mclane, *Land and Local Kingship*, p. 227.

[80] Kalyani Datta *Pinjare Boshiya* (Calcutta, 1996), pp. 47–60; Sambuddha Chakrabarti *Andare Antare: Unish Shatake Bangali Bhadramahila* (Calcutta, 1995), pp. 161–70.

households like that of the Nazims. An additional contribution could have been widespread homosexuality, the historical recovery of which is incomplete. Age-disparities could also have tilted the balance, with infant males and female slave-companions reared together from infancy, while the homogamous and later-acquired wives were much younger. The sonlessness of homogamous wives was elevated to something like a 'policy' across several generations in the ruling house of Bharatpur, when the Rani stated that from the time of

> Takoor Buddun Sing, the founder of the Bhurtpore family, and father of Maharajah Soorujmull, who died in the Christian year 1754, upto the present period of 8 successive Rajahs, no one has left a son born of a Rannee, but the successive heirs have all without exception been sons born of female slaves and *adopted* by the Ranee for the time.[81]

The conception and metaphors of motherhood held out by Manni Begam as legitimations of claims upon specific young males and females was widely shared in the political culture of the sub-continent. For as the claims above revealed, it was not birth but 'public acknowledgement' as 'mother' which was the effective determinant of the privileges of the women, as well as the relation of authority between the Rani and the incumbent ruler.

It is possible to speculate then, that both isogamous wives as well as some concubines within particular dynasties might have been forced to fail as birth-mothers, and compensated for this by purchasing and rearing children whom they then ceremonially 'adopted' as their own. Those women and girls who succeeded in the transition to motherhood may have had specific attributes that endeared them to the males, and their mothers. Those who had not, still served by rearing infants handed over to them. As one of the khawas in the regime of Mubarak-ud-daula petitioned, the Nawab had 'left with her an adopted son named Nasiruddeen Hyder... whom she nourished up from his Infancy'.[82] Concubines like Banni Khanum reared children like

[81] Memorial from the Rani of Bharatpur to the Chairman and Court of Directors, n.d., BC F/4/1350/53514. For similar claims by the Rani of Kalyan Singh, former naib Nazim of Bihar, that all the slave-concubines and their children were 'obedient' to her, *see* BPC, 9 May 1823, no. 158.

[82] Petition of Surphunnisa Begam to Charles Metcalfe, Gov-Gen. of India, n.d., BPC, 8 December 1835, no. 24.

Mubarakunnisa, who were not born of them, but of other women either in the household or located outside.[83] Thus even those who failed to become birth-mothers retained the possibility of social motherhood. In 1818, when, for the first time a detailed account of the de'orhis of the Mahalsarai was drawn up by the Agent of the Governor-General, it was found that a system existed of 'the concubines of departed Nazims... adopting Children, and thus introducing into the Nizamat persons totally unconnected with that Establishment'.[84] The person this referred to was one male slave, Babru, who was 'the son of a bhisti, but he calls himself the adopted son of the late Beebee Oojagur, a hurrum of Nabob Syefood Daulah'. Other older concubines of Mīr Ja'far, like Bi Phekun, were said to have adopted daughters. This pattern of childless women adopting daughters continued till at least 1849, when the deceased Bahu Begam left behind two such daughters.[85] Similarly, in Dacca, in 1849, it was reported that one Husaini Begam had reared 'two young girls named Pearee and Larly—who were left under her charge by former Nawabs.[86] Another such 'adopted child' of the Dacca household was initially 'made over' to the Nawab Ghaziuddīn when she was an infant by a 'wandering fakir'; the Nawab, on his death bed, transferred her to his concubine Sultani.[87] An identical case of a pilgrim having handed over his daughter for the *khidmat* (service) of the Nazim came to light in 1863; she had the title Malika i-Zamānī.[88]

[83] Mussamat Ruckee was the natural-mother of Mubarakunnisa, the daughter of Mubarak-ud-daula I, vide AGG Dale to Persian Sec., 8 March 1828, *MNLI*, I, p. 458 This daughter, Mubarakunnisa, also called 'Nawab Bhye', was 'brought up from her infancy with Bunne Khanum who treated her as a Daughter...' vide G. Swinton to AGG Ricketts, 23 October 1820, *MNLR*, II, p. 198.

[84] Enclosure 7, in Monckton to Lushington, 24 December 1817, BC F/4/618/15418.

[85] Statement exhibiting the names of old dependants of Bahu Begam, Extract GOB, GPP, 2 April 1857, no. 91, BC F/4/2708/194267.

[86] Collr. Dacca to R.H. Mytton, Commr. of Rev. 15th Divn., 5 September 1849, BC F/4/2396/128775. Pearee had been taken when two months old and placed under the charge of Hosseinee Begam and Larly had 'been maintained for ten months by Gazeeuddeen who on his deathbed, transferred her to Husaini.

[87] Mytton to Sec. GOB, 3 June 1850, and Collr. Dacca to Mytton, 15 May 1850, BC F/4/2433/133242.

[88] Note in evidence of Tarini Sankar, Prceedings of Commission of 1863, WBSA, GPP, April 1864, encl., no. 2.

The fact that the destinies of these sons and daughters were determined by the matriarch heads of the de'orhis, even though they were filiated to the Nazim or his cohorts as daughter or son, is significant for various reasons. Many such children were born of mothers who had been slaves of the specific matriarchs. An English visitor to early nineteenth century India had also commented on the 'custom in the zenana' whereby every lady had an 'adopted daughter' who was the child of a slave.[89] Bahu Begam, the first acquired consort of the Nazim Alijah, had been the daughter of a mughlāni in the de'orhi of Babbu Begam.[90] In the case of Alijah's three daughters, 'adopted' by Amīrunnisa (Dūlhin) Begam and the Bahu Begam, the natural-mothers had been the slaves of Walida Begam (Faizunnisa), and continued to serve the succeeding heads of the de'orhis. To quote the Agent

> Beebee Azeemun is the mother of the eldest daughter, and she, with her child, have for many years past lived in the Muhul Seraie of the Dhoolin Begum. The Begum... upon the death of her own child adopted this girl,... The Bhow begum never had any children by His late Highness, and to compensate for the disappointment, he committed to her care and adoption the two other daughters who with their mothers Beebee Lootfun & Beebee Zeentunissa have lived with the Begum since then, and are in every respect members of her family.[91]

This kind of patronage was based upon the ability of each head of a *mahal* to nurture and rear a pool of 'kinless' beings who owed allegiance to the head alone and constituted a part, as well as gauge of, the wealth and power of each head of de'orhi. As the memorial of one head of the Dacca mahalsarai put it in 1843, 'in the families of natives of rank especially in harems a number of females are maintained who have either lost their natural protectors or from other misfortunes are entitled to the commiseration of the charitable'.[92]

[89] Parks *Wanderings of a Pilgrim* I, p. 449. This was said of Colonel Gardner's household in Khasganj, near Agra, in 1835.

[90] Statement no. 4, BPC, 26 July 1822, no. 80.

[91] AGG Loch to A. Stirling, Persian Sec. to Govt., 14 August 1824, *MNLI*, I, pp. 306–9.

[92] Memorial of Hyatunnisa Begam, 14 December 1843, BPC, 26 December 1843, no. 32.

In the household of Muhammad Reza Khan's great grandson, Nawab Zafar Jang, a political officer described how:

> the mother of the son... was brought up from a very early age /4 or 5 years/ by the Mother of the Nawab expressly as a companion for her eldest son who was of the same age, the child born of her... lived always with his Grandmother.[93]

The girl referred to above had been 'purchased at Chitpoor [Calcutta]... for 12 Rupees from her mother', renamed as Rahiman and 'placed in the muhal as a common slave-girl'.[94] The grandmother, Nigarara Begam, subsequently claimed custody of all the children of the household, khawas-born or not. In a pattern repeated through the late eighteenth and nineteenth centuries, both apparently equal-status (homogamous) and other wives of the Nazims and the sahibzada were similarly young women either purchased by, or reared by, the matriarch of a de'orhi. The giving of various kind of 'daughters' in marriage by the matriarchs, in turn, structured patterns of kinship and power in the Nizamat.

Marriages of Daughters

Two significant aspects of the constitution of kinship in the Nizamat were a) the fact that all slave-born were filiated as 'sons' and 'daughters' of fathers, different from the patterns of filiation and status inheritance in British households in India at the same time which emphasized the inheritance of maternal status for slave-born and b) that many of these 'adopted' daughters were, as in the Nizamat, wives of the reigning men, their brothers and cousins. In addition, the concubines' daughters were important because of multiple political and social reasons. In the history of the house of the Nazims in Murshidabad (prior to the onset of Company rule) the men married to such daughters, the sons-in-law or *dāmāds*, apparently influenced succession, or could in turn pass it on to the sons born of such daughters. Succession to the masnad seems to have devolved through daughters. Murshid Quli Khan had favoured the succession of his daughter's son, Sarfaraz Khan, only to have the husband of his daughter (Shuja) intervene. Alivardi Khan,

[93] AGG Cobbe to Sec. GOB, 24 June 1835, BPC, 27 June 1835, no. 12.

[94] Statement submitted by N. Tahauwur Jang, 18 July 1835, BPC, Encl. 3, no. 31.

who usurped the masnad from Sarfaraz, in turn favoured the youngest daughter's son, whom he is said to have adopted as his successor as Siraj-ud-daulah. In turn, Siraj's accession was opposed by another of Alivardi's daughters, Ghasiti Begam. The successor to Siraj, Mīr Ja'far too was related to the Subahdar through his marriage with the 'half-sister' of Alivardi Khan, and had thus counted as a brother-in-law. In turn, he seems to have lived in some fear, and distrust of, Mīr Qasim, married to the daughter born of Ja'far's union with Shah Khanum. Yet she was by no means the only daughter in the household of Mīr Ja'far. Four decades later, upon the death of Manni Begam it was discovered that there had been at least two other daughters whose claims upon the Nizamat had been acknowledged by Manni Begam. Apart from Misri noted above, another daughter of Mīr Ja'far had been Khurshidunnisa, who was married to her paternal cousin (son of Ihtiram-ud-daula, brother of Mīr Ja'far): her grand-daughter in turn was nurtured in Manni Begam's de'orhi, and married to one of Mubarak-ud-daula's sons, Mīr Mehdi. Hence this double importance of daughters. In the Nizamat, they were the 'homogamous' wives of the sahibzada, nephews and grandsons in each generation living under the control either of the men's mothers or their own de'orhi heads and stepmothers; in time, they would not only govern the lives of their sons, but also as sisters determine the fates of their brothers' concubines. The reproduction of hierarchies within the de'orhis thus rested very crucially on these daughters of the household.

However, after the onset of Company rule, and its tightening grip upon succession to the masnad, the acquisition of grooms for such daughters appears to have lost its political edge. Instead of threatening to usurp the masnad, various male affines were recruited in order to extend the network of kin and clients of each de'orhi head alone. The existence of slave-concubinage in the core of the ruling group became specially critical at this juncture. The control of concubines' children and their selective disposal in marriage facilitated the expansion or consolidation of one's followership. Peirce has argued that in the context of the Ottoman dynasty, a standard feature was the making of the highest ranking members of the male slave elite into imperial dāmāds.[95] By thus enhancing the status and authority of the official,

[95] Leslie Peirce, *The Imperial Harem: Women and Sovereignty in the Ottoman Empire* (New York, 1993), p. 65.

dāmādhood created a strong bond of personal loyalty and indebtedness to the sultan and to the whole dynastic family. For Murshidabad, however, the determination of the status of these men—slaves or freedmen—is dependent entirely on vague ascription,[96] and on their upward mobility in the scale of stipends.[97] Some of those who counted as dāmāds, like Mirza Da'ud, an Ispahani merchant with whom Manni Begam appeared to have had certain building contracts, and Khalilullah Khan, came from outside Bengal.

Other dāmāds in the early part of the nineteenth century like Y'asubuddīn Ali Khan and Akbar Ali Khan Shamsher Jang appeared to have already been nephews or junior kin. Shamsher Jang was also a concubine-born son of the Rajmehal household (of Ihtirām-ud-daula, brother of Mīr Ja'far), with little independent wealth or property of his own.[98] The choice of these men as dāmāds may have been dictated by the same kind of criteria that governed the policy of concubinage for the Nazims; too well-connected a man may have endangered the balance of power. Furthermore, as the Nazims were pushed more and more into a fiscal corner, the impetus may also have been economic: as Mubarak-ud-daula said, it was preferable to get sons-in law 'from

[96] For instance, a man called Hakīmjang, whose wife was counted by the Agents in the group of 'grand daughters' of Mubarak-ud-daula (daughter of Salihatunnisa, aka Moti Begum, a daughter of Mubarak-ud-daula) was described as the 'Mootuburra' of Manni Begam, vide AGG Sotheby to Prinsep, 19 March 1823, BPC, 29 March 1823, no. 93. This is a mis-spelling of the term mutabarra', or freedman.

[97] In the case of Toraub Ali, the husband of Raz at us Sultan, this is implied because his name figures in the list of 1802 among the 'Shagird Pesha' on a very small stipend. This explains his struggle to actually receive the stipend of his 'wife' in 1808–11. About this attempt, Manni Begam was to say, 'before marriage he lived upon twenty six rupees, is it not very unreasonable .. that now instead of being satisfied with three hundred and fifty rupees... he claims four hundred more...' For further details *see* BC F/4/308/7064 and F/4/332/7633.

[98] For Shamsher Jang's antecedents, *see* AGG Cobbe to Macnaghten, 29 June 1833 in Extract Poll. Cons. of 11 July 1833, BC F/4/1475/57976. For two other dāmāds from the Rajmehal household, Taleb Hussain Khan and Y'asubuddīn Ali Khan, *see* Schedule I in C.E. Trevelyan to AGG Thoresby, 13 February 1834, *MNLR*, I, pp. 286–7. Taleb Hussain Khan, married to one of the daughters of Mubarak-ud-daulah I, 'Umdah Begam, was described as 'illegitimate' son of Abul Qasim, late Nawab of Rajemahal in AGG Loch to A. Stirling, 10 September 1824, BPC, 8 October 1824, no. 38.

families where he can avoid heavy expenses'.[99] Another Nazim said of one such dāmād, Himmait Ally Khan Bahadur Rustam Jang,

> he was not of very illustrious or distinguished Birth still his parentage was respectable and therefore my deceased father invited him from Benares to give him his niece in marriage daughter of his own sister and invested him with a high Title and settled upon him a Stipend of Rs. 1200 per mensem exceeding that granted to his sons by marriage with his own daughters above whom he also greatly distinguished Rustum Jung....[100]

Instead of a predominantly slave dāmādhood, the Nizamat then recruited such free poor men as dāmāds. These men were dislocated from their own regions (in Rustam Jang's case, from Benares; in another instance, from Dacca) and brought to Murshidabad for marrying the daughters of the house. This aspect of post-marital residentiality, of grooms in the house of the wife's family, identified by sociologists as uxorilocality, remained characteristic of the lives of most of the sons-in-law of the early Nazims. For instance, a daughter of the Nazim Mubarak-ud-daula I, Badrunnisā lived with her husband Shams-ud-daula in the palace, till the latter was convicted of conspiring to overthrow the East India Company in 1799, exiled from Murshidabad and allowed to reside in Dacca,[101] from where eleven years later he pleaded to have his wife sent to him. On this occasion, the Agent noted that it would be the first time that a woman of the family had left the *killah*.[102]

Significantly, there were two other aspects of the treatment of dāmāds that argued for the choice of less powerful men: one was the apparent suspicion that attached to these grooms, since guarantees were sought to be procured by the matriarchs from these men through 'marriage settlements' promising extraordinary sums of compensation as mahr should these grooms set up alternative lineages.[103] In other

[99] *CPC*, IX, no. 1059, p. 232.

[100] Translation of a letter from the Nawab, 16 October 1808, *MNLI*, I, pp. 97–8.

[101] Extract BJC, 18 December 1800, Progs. of trial of Shams-ud-daula and others, BC F/4/128/2371; Nandalal Chatterji, 'Shamsuddaulah's Intrigues Against the English', *BPP*, 53, 1, 1937, pp. 31–4; and N.K. Sinha, 'The Case of Mirza Jan Tuppiṣh: A Treason Trial of 1800', *BPP*, 72, 1953, pp. 39–42.

[102] Extract Poll.letter from Bengal, 2 February 1808, BC F/4/247/5567.

[103] Petition of Nurunnisa Begam, daughter of Mubarak-ud-daula, to Governor General in Council, 29 May 1837 in BPC, 4 July 1837, no. 21. She too was taken 'as

words, though Nazims signed no *kābinnāmahs,* these lesser males did: perhaps this difference underscored the lower standing and dignity of the males involved. The second was the established practice for the Nazim to reclaim the marriage portions of the females of his family. The near-adrogation that was implicit in these marriages of the first half of the nineteenth century then further enhanced the need each de'orhi and each Nazim felt to establish immaculate claims upon the deference of these grooms.

While uxorilocal grooms increasingly came from the group of males represented as the 'cousins' of their brides, the patronage of the Company may have had something to do with this inward turn in the acquisition of grooms. Just as women were made Begams by the favour of the Nazim, certain men became dāmāds by virtue not of a particular sacralised ceremony but by virtue of the *dāmād-i khil'at* bestowed on them. The value of these rose at the end of the eighteenth century when the Company launched its financial reorganization of the Nizamat. In the accounts submitted by Manik Chand, the Diwan of Manni Begam, to the Committee in 1801, it is said that 'the distribution of khillauts occasioned at least 7000 Rupees each time'.[104] Both the expenditure incurred, as well as the political implications of such grants, worried the Agents and the Government of Bengal. One way of regulating both was the device of setting up of a fund (from deductions from the sixteen lakhs annuity) for the marriages of daughters, a device first thought up in 1796. By the 1820s, this money was in turn being used by the Agents and the Nizamat stipendiaries alike to fund the marriages of numerous daughters, many of whom were described by the Agent either as 'adopted' or as 'illegitimate'.[105]

---

an infant' into the de'orhi of Babbu Begam and married to a man called Mīr Mohsin Salabat Jang. She claimed that according to the practise of 'our family' and her 'eleven other sisters', the groom had solemnly promised to 'abstain from illicit intercourse with other women on pain of forfeiture... as guarantee for the faithful discharge of this and other obligation as a husband he settled on me a marriage portion amounting to seven lakhs of rupees claimable by me'. A study of the total of four petitions that were given in by her reveal that the groom and she had separated at least twenty-five years before this date, during the reign of Nazim Nasir-ul-mulk Babar Jang.

[104] Letter of Pattle, Rocke, Hayes, to Secy. to Govt. in OIOC, B.C. F/4/249/5601.

[105] Enclosure 1, AGG W. Loch to Pers. Sec. to Govt., 30 August 1824, *MNLI*, I, pp. 317–21. A scrutiny of this list, compared with evidence of other files, reveals that

Till 1836 such grants of money were mainly given for marriage-expenses to the elders, rather than as 'dowry' for the daughter. Thus it may be possible that in a cash-starved de'orhi, one of the important ways of consolidating its resources was in fact to arrange for an inter-marriage between various members under its control. Hence, what appeared as marriages between cousins from the 1820s became significant since both de'orhi heads and Nazims sought to retain control of these in their own hands.

In 1824, Nazim Walajah, had begun to 'prepare for the marriages of his three neices, the daughters of his late brother', Nazim Alijah. The contest over what superficially appeared to have been an avuncular and patriarchal privilege, was actually a battle for the control over female children between the ages of 8 and 14 who had been born to the slave concubines of the previous Nazim.[106] The Begams (Bahu and Amīrun, the deceased Alijah's consorts) claimed that both precedent and right was on their side in the matter of the marriages of these girls: not only had all such daughters of previous Nazims been bestowed by the heads of the mahalsarai, but also that this right had been 'confirmed to them by the act of their husband who had

---

the apparent 'cross-cousin' marriages that were being funded were unions between various concubines' children, male and female. Thus in one instance, the daughter of Mīr Mangli is proposed to be married to the son of 'Umda Begam. Both Mangli and 'Umda are normally referred to as 'son' and the 'youngest daughter' of Mubarak-ud-daula I respectively. Like Mangli, born of a slave concubine, 'Umda too was born of the concubine Brijmahābakhsh, and was reared in the mahal of Babbu Begam vide AGG Russell to H.T.Prinsep, 23 June 1821, BPC, 7 July 1821, no. 76. 'Umda's marriage with Taleb Hussain Khan brought into the kingroup Taleb Hussain Khan's other slave-born children, for which *see* AGG Ricketts to Persian Sec., 21 Nov. 1822, BPC, 7 December 1822, nos. 56–7, apart from the son born of 'Umdah called Wahed Hussain Khan. Three of these daughters, born of Taleb's concubine Lutfunnisa, were then married into the Nizamat, as revealed in Loch's list of 1824. A similar deduction follows from the consideration of other names on this list, like that of the daughter brought up by Banni Khanum, Mubarakunnisa. The latter's marriage with Y'asūb-ud-dīn extended the kingroup with Y'asūb-ud-dīn's slave-born children as well as his children with Mubarakunnisa. Another daughter of Mubarak-ud-daula I, Rahīmunnisa, had her son Bakr Ali marry Alijah's concubine-born daughter, Vazīrunnisa .

[106] AGG Loch to A. Stirling, Persian Sec. to Govt., 14 August 1824, *MNLI*, I, pp. 306–9.

committed those children to their care'. The Agent supported the Begams partly because of the claims of traditionalism, and partly because he suspected that the Nazim was using the pretext of 'marriage portions of his nieces' to demand three lakhs of rupees from the Government. Insufficient as an explanation for the desire to control such slave-born daughters, it was the wider strategies of clientage-creation that could provide a better insight into a direct political investment in such daughters.

Bestowing a wife was simultaneously a political transaction reinforcing or establishing deference and acceptance of the bride-giver by the bride-receiver, and an economic transaction, related to the protection afforded in law to the financial estates of wives. The theoretical principle was that the mahr (dower) paid by the groom became the property of the wife. By extension, if such a wife was also the purchased/kinless or slave-born adoptee of the Begams or of the Nazims, the wealth that accrued to her in 'dower settlement' came directly to the coffers of the bride-giver. In addition, particularly where there were children expected to be born of this marriage, they would elevate the status of the de'orhi heads as 'grandmothers', and further strengthen their authority over incumbent generations of Nazims and sahibzada, as well as establish fiscal claims in the wealth and holdings of the state as a share in the 'patrimony'. For instance, there was the childless Amīrunnisa, whose 'gift' of Alijah's concubine-born daughter to a young Nazim (Humayunjah) in 1826 finally achieved for her the motherhood of a male that was a prerequisite to authority over other women. Both first-generation acquisitions and second-generation *assimilés* were important precisely because their natural-mothers were not those from whom such progeny derived their filiation or status. Their reproduction as non-slave 'sons' and 'daughters' of the lineage depended instead on the matriarch head's command over material resources. In turn, this material investment in a male or female infant, while socially reproducing this child as a son or daughter, was also expected to reap a wealth of allegiance, service and returns in the future, as by the marriages of such sons and daughters, networks of clients and dependants could be arranged for the strengthening of particular factions within the Nizamat. Thus both the mahal heads and the Nazims tried to expand their resources of human and inanimate wealth by manipulating the marriages of such daughters.[107] Noticeably, the slave-women who had borne these daughters were neither the initiators of such 'alliances', nor the beneficiaries of the dower claims upon the estates of the dāmāds.

The Limits of Incorporation

The attempt to create clients, rather than permanent and equal-status allies thus undercut the marriages of all daughters. For instance, the grand-daughter of Mubarak-ud-daula married her ailing cousin (the son of Nawab Jān, another son of the deceased Nazim); her uncle and father-in-law, Nawab Jān, took away all her jewels and sold and pawned them.[108] Given that every half-brother's hand was also turned against the other, such marriages between cross-cousins could neither strenghthen affect between male affines, nor consolidate the material resources and funds of both de'orhis. Hence, dāmāds like Salābat Jang were physically assaulted by his wife's half-brothers for 'some deficiency in the ceremonial of salutation'.[109] Moreover, both the Ispahani dāmāds and the Murshidabad-based grooms had very tenuous relationships with those 'wives' who were daughters in the Nizamat: death and the swings of political fortunes ensured quick separations. In at least two instances in the early nineteenth century, and once later, the death of one such wife was followed by remarriage with another of the daughters of the Nizamat, leading very often to bitter disputes between rival factions over the dower and inheritance claims when a dāmād died and the daughter became a widow.[110]

[107] For the conflict over the choice of grooms for three of Alijah's daughters, Bahu Begam choosing men born of women in her faction, and Amīrunnisa choosing grooms from her faction, *see* Poll. Letter from Bengal 3 October 1829, BC F/4/1228/40157.

[108] Suptdt. of Nizt. Affairs to Pers. Sec., 25 October 1813, BPC, 6 July 1816, no. 62; AGG Russell to Ch. Sec. GOB, 1 April 1822, BPC, 10 May 1822, no. 92.

[109] AGG T. Brooke to Ch. Sec. to Govt., 20 April 1812, Extract Poll. Cons. 1 May 1812, BC F/4/401/10094.

[110] *See* Chapter 4 for a discussion of disputes over the property and stipend of Shahamat Jang, dāmād through marriages with Afzalunnisa and Rahīmunnisa Begam, BPC, 10 October 1809, nos 54–9, BC F/4/312/7131; and over the estate of Rustam Jang dāmād through marriages with Shamsuddī Begam and Shāhrukh Begam, BPC,

The patterns of status inheritance (children of the lineage) for the concubines' daughters in the first generation ensured a subtle manipulation of reproductive strategies by the grooms involved. Since their brides moved from the control of one matriarch to the domain of another matriarch and children born of these unions were not reared by the grooms but by the matriarchs of the particular de'orhis, each male who took such a daughter as wife simultaneously created a lineage of his own through a 'third' force—the contracting or buying of slave women from men and women not amenable to the matriarchs. (*See* Appendix I). The bitter protests of many elders—matriarchs and patriarchs alike—against infringements of their rights to provide 'wives' for the junior males in the household were significant in this context.[111] Such infringements could occur because of the machinations of other courtiers, officials or servants contracting their women (slaves and non-slaves alike) to an incumbent ruling male. In the household of the Naib Nazims of Dacca, for instance, an incumbent Nawab complained against his heir-apparent for having disregarded his isogamous wife (paternal cousin) and been induced to marry the daughter of his syce-groom.[112] Similarly, Badrunnisā, the grandmother of the Nawab Ghaziuddīn of Dacca complained against two men who had persuaded the Nawab 'to marry and to settle, in addition to large dowries secured to them by written documents, handsome pensions for their support... to marry a low woman... and to settle an allowance of 1500 Rupees per month for her maintenance [which]

---

15 July 1809, nos 63–6, 3 October 1809, nos 44–6 and BC F/4/304/7006 and F/4/333/7635. Similarly in 1863, the dāmād Zainulābdīn, who was also a grandson of Mīr Mangli, married two daughters of the Nazim Faridūnjah consecutively: his first wife was Fātima Begam, daughter by Bībī Faizunnisa Khanum, mamdūd, vide AGG Lt. Col. Thomson to Sec. GOB, 7 December, 1863, and 22 December 1863, WBSA, GPP, February 1864, nos 1 and 4.

[111] Translated letter from Najībunnisa Begam of 29 Rabī-us Sanī 1247 Hijri, BC F/4/1456/57367. Her claim that it was not customary to acknowledge as 'sons' those born outside the *mehal* was substantiated by Darāb Ali Khan in 1836 who said 'the nazim's mother was opposed to her son's having affairs with women out of the *zunanah*'.

[112] Nawab of Dacca to Commr. Dacca Divn., 2 May 1830, BC F/4/1343; Commr. of Dacca Divn. to Sec. to Govt. Persian Deptt., 5 May 1830, referred to the father of the girl as 'a menial servant in attendance on the young Nawab's person, styled Daroga of the stables' and to the ceremony being 'secret'.

they punctually receive as soon as the stipend is paid out of the Collector's treasury'.[113]

The complaints revealed that while slave-marriage/proximity continued to produce isogamous cousins over one or more generations, the availability of slaves outside the household continued to provide an alternative lineage-making strategy for the males of the same household. For this reason, by the thirties and forties a characteristic pattern developed in the lives of both affines and consanguines, dāmāds and sons—a pattern of demographic failure, or delimitation, on the part of their isogamous wives, coupled with sexual reproduction from freshly acquired slave-wives or older slave-concubines. The lives of both Mubarak-ud daula's and Alijah's daughters bear testimony to this process. Amongst the older generation of daughters, one married to Taleb Hussain Khan, found her conjugal home ruled by her husband's *mankuha* 'who ruled the other females of his family including the Begam with a *rod* of iron, nothing transacted but through her means'.[114] Another daughter wedded to Saiyid Sikandar Ali had only one daughter with her husband; he had other sons and daughters from his slave-concubines, and died in debt 'due to Sultana and other persons'.[115] Similarly, one of the three daughters of Alijah, Vazīrunnisā, was married to a relative who had had two sons with a 'haram' of her father already.[116] Like her cohort Hosseini, Vazirun too appeared to have had borne no children of her own as a wife. Even more obvious were the instances of the two 'cousin' wives of the Nazims between 1810 and 1845. Alijah's homogamous wife Amirun (born of Faizunnisā/Walida Begam's daughter Sāhibzadī Begam, and hence 'first cousin' of Nazim Alijah,) produced no living child, but he left three

[113] Badrunnisa Begam to Dy. Gov. of Bengal, 18 June 1838, BPC, 18 July 1838, no. 6. For one of these women who claimed a 'nicca' marriage with the Nawab, and the status of widow after his death, *see* petitions of Goolzar Begam, 4 February 1844 and 25 May 1846, BPubC, 3 June 1846, nos 81–4. These multiple marriages clearly had non-reproductive significance, in the light of later suggestions in the historiography that this Nawab was bisexual, or at least dressed like an 'English lady', cf. Rafiqul Islam, *Daccar Katha*, p. 110.

[114] AGG to Pers. Sec., 21 November 1822, BPC, 7 December 1822, no. 56.

[115] Trans. letter from Jigri Begam, recd. 22 February 1811 and Pers. Sec. GOB to Suptdt. of Nizt. Affairs, 5 July 1811, BC F/4/372/9259.

[116] AGG Dale to Pers. Sec., 31 May 1828, and Pers. Sec. to AGG, 16 August 1828, BC F/4/1228/40157.

daughters by other concubines. One of these daughters, 'Umdatunnisā or Khurshid Mahal, born in a slave-servant's (*chabūksawār*) house, was in turn married under the auspices of Amīrun, to another Nazim, Humayunjah in 1826—this marriage too produced no children. Humayunjah's son and successor to the masnad was born of yet another slave-concubine, albeit a hired one.[117] Later in the nineteenth century, even when such homogamous wives of Nazims did successfully reproduce members of the lineage, as Shams-i-jahan did, some mode of ensuring the usual limit of motherhood persisted: of eighteen children she had borne, only one son and two daughters survived.[118]

Most of the affect and authority sought to be established by the marriages of such daughters thus appeared ephemeral at best. Amīrun's gift of a bride to the Nazim Humayunjah failed to prevent the latter's attempts to humiliate her, when upon reaching majority, he assumed the reins of government.[119] Nor did Bahu Begam's gift of her 'reared' daughter yield much more: Hosseini, the concubine-born daughter of Alijah, who had been married to another grandson of Mubarak-ud-daula in 1829, was widowed shortly after in 1833. The story of this particular daughter of a Nazim illustrated yet again the dangers of being fully incorporated into the lineage as well as its limits. Two years into her widowhood, Hosseini's own concubine-mother (Bībī Lutfun) died, and the Agent apparently upheld the daughter's right of inheriting the deceased mother's property.[120] Within six

[117] Two different reports exist to substantiate this: one by the AGG, reporting the birth of a son to Humayunjah on 29 October 1829, which meant that he had 'three children alive—namely a son and two daughters... all the offspring of concubines' in Dale to Stirling, 15 November 1829, BPC, 20 November 1829, no. 51. The birth of this son had followed the death, in 1828, of another son, born presumably of another concubine, since AGG Melville had refused to pay the correct condolence on this occasion because the child was 'illegitimate'. *See* Melville to Stirling, n. d. [circa 1828] BC F/4/1228/40156. Neither officer mentioned the 'homogamous' wife of Humayunjah as the mother. The details of the birth of Faridūnjah were the cause of the investigations of 1836, IPP, 1 August nos 85–7, and nos 90–3.

[118] Memorial of N. Shumsijahan Begam to Lt. Gov. of Bengal, n. d., WBSA, Judcl., November 1893, no. 56.

[119] For reports of Humayunjah's use of mimicry, ribald songs and exhibitions to taunt his 'mother-in-law', *see* AGG Cobbe to Sec. Macnaghten, 17 June 1833, and Amīrunnisa to Lord Bentinck, 28 June 1833, BC F/4/1522/60090A.

[120] AGG Cobbe to Nawab Nazim, 22 April 1835, *MNLI*, II, p. 6.

months, this twenty-one year-old daughter stood accused of 'exposing herself in a cheep on the river'.[121] Another widow, Shahzadi Begam, also born of a concubine, was accused of 'conduct... the most barefaced... a public scandal to the whole family' after she tried to lease and mortgage the inheritance of her young son. The alacrity with which specific kin, male and female alike, combined with British officials to accuse such young women of sexual misconduct revealed yet again the cleavages within kinship structures into which incorporated kin also fell. In both the instances above, such charges were preliminaries to the denial of their inheritance claims as both daughter and wife.

The real symbolic charge in such accusations was that of faithlessness. Occurring as they did precisely after the death of a deceased 'husband/master', they obscured the fact that post-mortem manumission to a slave-concubine, especially if she had borne a child, was both morally enjoined as well as practised in fact. However, *this* particular gift of 'freedom' could be confirmed only by the dead man's heirs. In the increasing fiscal embarrassment in the Nizamat, perhaps these charges were meant to initiate a negotiation by heirs of deceased males, to indemnify themselves for this 'gift'. But even more significant, since slave-concubines were accepted as near 'extensions' of the master/husband's persona, the 'freeing' of the slave at the death of the master meant that he was resurrected in the living person of his favourite. This was both a source of anxiety to the surviving heirs of a dead man, as well as cause for the renewed obligation of the slave-wife or concubine to be as inviolable as the master had been in his lifetime. Thus charges of 'exposure' of the person of the concubine and wife, after they had become mothers of the lineage, and after they had been 'widowed', was to charge these women with a grave lapse of obligation to the master, to his lineage, and to his heirs. Using these charges against specific women to diminish their stipends, or erase their names altogether from the record books was, as can be expected, conducive to a fiscally straitened Nizamat; it was apparently another way of denying her place in the inheritance of deceased members. The Agents were happy to go along with such punitive measures, and to reinforce

[121] AGG Cobbe to Sec.to GOB, Macnaghten, 14 October 1835, BPC, 27 October 1835, no. 27; BC F/4/1168/30773.

them wherever they could, in the interests of the 'observance of common decorum'.[122]

As if pecuniary reprisals were not adequate for curing this so-called 'notorious profligacy and disregard of appearances' among the female branches of the Nizamat, other methods were devised by English officials as prophylaxis.[123] Convinced that the 'natural and most suitable guardian of a woman's honour' was a husband, an Agent in 1836 initiated a policy of getting all such widows 're-married'. In a repetition of earlier efforts by Loch to get the 'illegitimate' daughters married, this Agent set about arranging matrimonial alliances for 'widows'. The Agent had discovered

> a custom of this family... that the widows never remarry... finding this usage does not obtain in society generally and that it derives no sort of sanction from the Mahomedan Law, I... quietly consult[ed] His Highness on the propriety of reverting to the Law, instead of abiding by the practice.[124]

Re-establishing the authority of 'husbands', male cousins and 'sons' according to some grand patriarchal design of domesticity only exchanged one form of conflict for another. There were enough indications that those who inherited control over these concubines did hire such women out in turn to different men, depending on the direness of the straits they were in. This was as true for women who were metaphorically 'mothers' of an heir (but in reality the slave-women or *kanizān* of deceased males) as it was of 'widows' and 'daughters'.[125]

[122] AGG F. V. Raper to Secy. to GOB, F. J. Halliday, 1 November 1844, BPubC, 8 January 1845, no. 51. A similar pecuniary punishment was meted out to Bismillah Begam a 'haram' of the deceased Syud Wahed Ali in 1852, vide Letter from GOB to Court of Directors, 24 June 1852, BC F/4/2476/138839.

[123] AGG Melville to Macnaghten, 20 June 1836, BPC, 28 June 1836, no. 2.

[124] Ibid.

[125] For an instance of the former, *see* AGG to Sec. to GOB, 28 November 1835, BPC, 8 December 1835, no. 18. In this the son of Mīr Mehdi, a youth called Mubarak Ali, was punished by the Nazim for having taken the 'hurm of his father' with him on a boat, and having her visited by a servant of the Nazim's, the Darogha of the Shutarkhana. For an instance of the latter, *see* Commr. of Dacca to Ch. Sec. Swinton, 14 January 1832, BC F/4/1432/56565. According to this, some men of Dacca had complained against a woman called Kaniz Fatimah, who had been 'widowed', had been taken as a 'mistress of a Hindoo merchant of property' and was subsequently said to have been 'little better than a common prostitute'.

CALIBRATING LEGITIMACY, COMPROMISING CHANGE

Myriad contradictory pulls developed within each generation of slave-born children and slave-wives, making the structure of each such household fissiparous, and each unit affectively fragile. By the 1850s, the failure of marital strategies which made affinal males (brothers-in-law and sons-in-law) indistinguishable from consanguinal males (the half-brothers, cousins and uncles) and foster-sisters and female cousins inseparable from wives was spoken of by a Nazim educated under English tutors and eager for change. In the attempt to reinvigorate the divisions, the Nazim Faridūnjah spurned the previous policy of cross-cousin marriages within the Nizamat in terms of distaste

> such alliances where the parties spring from the same blood, are unsuitable and unbecoming... counter to the feelings of propriety and moral conviction.[126]

The Nazim also revealed that such marriages had reinforced the acute pecuniary distress of all concerned. The example he gave was that of his own sister, Gītiāra Begam, married in 1836, while very young, into an impoverished but collateral branch of the family.[127] To improve both the material and socio-political contours of the family, the Nazim initiated a distinct change in the acquisition of spouses by attempting to marry two of his own daughters[128] to the two sons of the Vazir of Awadh.[129] Ironically, since such marriages also required appropriate displays of wealth, and precedents existed for grants of about a lakh

[126] Trans. Nawab Nazim to AGG, 1 April 1857, BC F/4/2708/194269.

[127] AGG to Sec. GOB, 6 May 1836, BPC, 12 May 1836, no. 3. The groom was Ali Muhammad Khan, and the titles given to him as a dāmād were Iftikhar-ud-daula Basalat Jang. As a son of Sultan Ibrāhīm, he was counted among the grandsons of Khalīlullah Khan.

[128] One of the daughters was eleven-year-old Sultanunnisa, daughter of Malika Zamani Begam, cf. AGG to Offg. Sec. to GOB, 22 November 1862, WBSA GPP, May 1863, no. 21. The other appears to have been ten-year-old Padsha Begam, daughter of Shamsijahan.

[129] See correspondence on the marriage of the daughters in BC F/4/2708/194269. The last King of Awadh, Wajid Ali Shah (1847–56) had also married a daughter of this Vazir, Ali Naqi Khan, *see* Fisher, 'Political Marriage Alliances', *CSSH* 25, 1983, p. 614. For the Nazims in Bengal to thus give daughters to this family was an important move. For subsequent accounts of expenses for the marriages of the daughters, and *dāmādi khilats*, *see* BPP, December 1874, nos 1–2; February 1877, nos 3–4, April 1877, nos1–3 and January 1878, nos 28–9.

of rupees for each daughter of the Nazim (given in the case of Ùmdatunnisā and Gītiārā), both the Agents and the Government of Bengal tried to establish a scale of grants which would diminish according to the birth and origin of particular daughters. AGG Thomson worried that the Nazim had fourteen unmarried daughters, 'ten of whom are born of mumtooahs or kept women, for whose settlement it can scarcely be the intention of Government to make such large grants'.[130] His superiors, while rehearsing the well-worn axiom that 'daughters by wives and daughters by concubines are equally legitimate' proceeded to establish a diminished scale (Rs 20,000) for those daughters who were lesser-born, and a higher sum (Rs. 40,000) for the others.[131] Progressively, over the next two decades, while the Government of Bengal tried to whittle down these sums further, making them largely account-book transactions,[132] these differentiated sums in turn acquired symbolic value. Hence, for a son of a Nazim to accept Rs 20,000 (instead of 40,000) as marriage-allowance for his sister was tantamount to accepting inferiority of status for himself. Thus Ali Kadr, the first-born son of the Nazim, wrote about a list of eleven daughters of the Nazim (each of whom had been granted the lesser sum) that if he were to accept this amount for his 'own sister' Malikunnisa urf Hurmuzi Begam, he 'would be seriously compromising her rank and sacrificing her status in the family'.[133]

As the numbers of the Nazim's sons and daughters progressively increased in inverse proportion to the control of the elders and the matriarchs, (and included children from two Englishwomen whom the Nazim espoused on a sojourn between 1869 and 1881), successive investigations and commissions of enquiry thus yielded many such daughters and sons. The only difference between the thirties and the seventies was that by the seventies, Agents and other officers of Government had both been been trained in textual law, and had begun to cite particular judicial decrees of the High Court in Bengal (and

[130] AGG, Maj. Thomson, to Offg. Sec. to GOB, 22 November 1862, WBSA, GPP, May 1863, no. 21.

[131] Sec. GOI, Foreign, to Sec. GOB, 21 March 1863, ibid., no. 34.

[132] R.L. Mangles, Offg. Sec. GOB to Sec. GOI, Foreign, 9 February 1877, BPP, February 1877, no. 4.

[133] Nawab Ali Kadr Syud Hussain Ali Mirza Bahadur to AGG, 4 April 1877, BPP, January 1878, encl., nos 28–9.

the Privy Council) regarding 'legitimacy' in Islamic law. Hence, one such Agent pointed to both to urge that 'the legitimacy, or legitimation, of a child of Mohammedan parents may properly be presumed, or inferred from circumstances without proof, or at least any direct proof, of marriage'.[134] By 1878, the Government of Bengal had agreed to dispense with 'proof' of the canonical variety, since

> no actual proof of *nikah* is forthcoming, but the Nawab Nazim's declaration that she was his *nikah* wife should determine the point... when the forms of marriage are so indefinite and the distinction between formal and informal marriages so slight... it is better to accept the views of the head of the house on the subject.[135]

Accepting the word of the 'head of the house' was not the result of a greater understanding of the political rituals incorporating slave-women and their children into the Nizamat. It was a new bargain. A nuanced understanding of these processes would have prevented Agents from asking for proof of 'ceremony... in the marriage whether shadee or nickah *with each* woman' as many of them had continued to do in the late sixties and seventies.[136] The doing away with 'proof' of marriage in 1878 was offered as an olive branch at a specific point in the relations of the Government of India and the Nizamat. From 1873, the Government of Bengal had instituted a Committee to investigate and repay the debts of an absent Nazim, which amounted to thirty lakhs. In the round of reorganization of lands, funds and stipends that accrued, the Nazim was to receive only the amount of what was called 'his pocket-money'.[137] The cooperation of the oldest among the sons, Ali Kadr, in gathering the necessary information, and in managing the new system, was vital. A range of political and financial compromises thus determined that slave-born or not, Ali

[134] Offg. AGG, G. Hodgkinson to Sec. GOB, 31 July 1877, ibid.

[135] H.C. Cockerell, Offg. Sec. to GOB, to Sec. to GOI, Foreign Dept., 31 December 1877, BPP, January 1878, no. 30.

[136] AGG Buckle to Walah Kadr Syud Hosain Ali Mirza, 13 March 1869, *MNLI*, II, p. 648; for a denial of patrimonial claims on a paltry sum of Rs 200 on grounds of 'illegitimacy' of the mother, *see* AGG Buckle to Sec. to GOB, 31 August 1868, BPP, January 1869, no. 67 and genealogical trees enclosed in nos 71–2.

[137] Viceroy and Gov.-Gen.-in-Council to Sec. of State for India, 4 August 1873, BPP, January 1874, no. 72.

Kadr's claim as successor to his father was to be recognized rather than that of the much younger Iskandar Ali Mirza, born of Shamsijahan, the isogamous wife, married as she put it by shādi.[138] In addition, too fine a point could not be put on the kinds of contracts a Nazim made while in England, since there were now English-born sons and daughters whose interests would be adversely affected by such refinements. Even as it nailed the coffin of the Nizamat, the indenture of 1880 thus guarded the interests of his four English-born sons and daughters, without a word about the kind of marriage-ritual they had been born of. The other children were left to battle their eldest brother for their shares, as we will see in the next chapter.

[138] Memorial of Shumsijahan Begam to Lt. Gov. of Bengal, n.d., WBSA, Judcl. November 1893, no. 46. According to her, the Nazim was 'never married to any one else'.

4

Slaves, Wealth and Inheritance

Joseph Miller studying what he called a 'political economy of followers'[1] outlined a polity that stressed people as sources of productivity, so that becoming economically wealthy and politically powerful meant aggregating human dependents of different kinds. Material goods were critical to this economy of deference in that their transfers became a mediating procedure for acquiring this deference; according to standards of patronage, gifts and largesse, these goods were redistributed among dependents. Such transactions confirmed the connotations of superiority and inferiority when material goods changed hands, and reinforced the desire of every subordinated recipient to avoid having to resort to such transactions whenever possible. This kind of political economy of deference and honour, mediated through the gifts of money, cloth and food, was characteristic of the Nizamat too. As Mubarak-ud-daula I, finding that some of the stipends he was supposed to disburse had been separated from his own, explicated

> Kings, Vazirs and other Chiefs are of one species of men their Degrees of honor and Dignity arise solely from... their power to do good... according to their degrees of power to benefit mankind they are denominated Viziers Amirs &ca otherwise they are all on a footing and the wealth of crores can make no difference. Their dispensing it constitutes their Dignity.[2]

[1] Joseph C. Miller, *Way of Death: Merchant Capitalism and the Angolan Slave Trade, 1730–1830,* (London, 1988).

[2] From Mubarak-ud-daula, recd. 15 November 1790, BPubC, 3 December 1790, no. 7.

By his reasoning, political and social stratification were functions of the gift. The separation of individual pensions and stipends from the control of the Nazim by undermining this power of *giving*, destroyed the very basis of his dignity. Instead of dependence and obedience that all followers and relations owed to him as a result of being recipients of his beneficence, they would be elevated to the ranks of 'sharers' and hence his equals. Perhaps no better record of the congruence of honour, public acknowledgement of authority, and the power of the gift exists in the records of the Nizamat than the ones generated in the course of a contest between the Nazim and the Walida Begam in 1811, over the issue of *who* had the privilege of the 'gift'. The substance of the contest were the stipends of the 'Officers and dependents' of the various de'orhis and the 'annual presents' of *zamistānī* (winter clothing) and other 'customary gifts' to the sons and daughters of the Nazims. Elderly Wālida Begam, Faizunnisā, claimed that these were hers to give by virtue of the position she had succeeded to, as the Head of the Mahalsarāi, and the seniormost woman of rank in the Nizamat. The Nazim claiming the greater splendour for his consort, argued that the only way of ensuring this was through the distribution of stipends and allowances, for then alone 'all the principal dependants, inferior servants, domestics and others attend at this de'orhi and thereby give it an appearance of consequence and importance, and perform the various offices connected with it'.[3] This hieroglyphic representation of honour and dishonour through the visibility of giving and receiving was not limited to the Nazim alone. A male descendant of a collateral branch of the house, Da'ud Ali Khan objected to the 'stipends' of the old dependants of the deceased head of the family being paid by the head of a rival branch thus:

'these are my domestics who have acted in their several capacities and duties in regular succession; and continue in attendance on me. When therefore they are excluded from the list of my own establishment and are provided for under that of the Nuwab Mudarood Dowlah, how can they continue their services to me?... I cannot on any account consent that the whole allowances should be distributed by the Nuwwab's people'.[4]

[3] Zainuddin Ali Khan to Rocke, n.d., BC F/4/372/9260.

[4] Daood Ullee Khan to Persian Sec. to Govt., recd. 24 December 1821 in *MNLR*, I, p. 200.

This was hardly a bid for sizeable manpower, (the number involved were six 'servants'), or for sizeable income, (the total sum in question seemed to have been Rs 189 only). The claim can only be understood in the context of the authority derived from bestowing. This was just as true for transactions *within* a group of kin, as it was for transactions between kin and non-kin. The son of Muhammad Reza Khan, Dilawar Jang, had made this clear to a Governor-General who puzzled over why the former did not disburse the sum allotted for his brother, Behram Jang's, 'widow'. Abjuring all mercenary motives, Dilawar Jang said,

> if I am only to be the medium of conveying to her the Pension, without possessing any authority over her, I shall be placed in the situation of a Khazanchee....[5]

Since power over people was so clearly associated with the gift, the the status of the recipient was *ipso facto* that of a dependent and junior; no inversion of this code was free of dishonour. Wālida Begam herself pronounced upon this one 'Id, when the young Nazim, instead of presenting his *nazar* in person to her, sent it through his eunuch, Siddi Javed; the gesture, she claimed, was a declaration of 'her inferior rank and consequence' not only because she had been denied the personal homage of the Nazim, but because it made her appear to be at the receiving end of a transaction from the eunuch. A very similar statement from a granddaughter of the Nawabs of Dacca (two brothers who succeeded each other in the period 1785–1831) explained that she, the elder in age and senior in rank, could not receive her allowance by the 'hand of Hyattunnisa Begam as... she was a Bondmaid of my late uncle'.[6] To receive sustenance or allowance from the hands of a slave was felt to be an inversion of mastery precisely because moral authority and power over people rested on giving.

The disbursement of money, stipends and other such allowances was thus an aspect of authority distinguishing the elder from the junior. The various sums, referred to as wages/ stipends/ salaries/ allowances

[5] Dilawar Jang to Persian Translator in Edmonstone to Pattle, 1 December 1798, *MNLR*, I, pp. 43–4.

[6] Petition of Muzzeedun Nissa Begam (Potee) to Dy. Gov. of Bengal, 25 March 1838, BPC, 25 April 1838, no. 19.

in English, thus followed from a prior relationship of authority and acted as advances critical to initiating, and retaining, services within a relationship of dependence and deference. Within this relationship other distinctions like personal affection, or individual characteristics and skills then determined the amounts each received. Part of a very intricately related economy, these stipends and allowances differed in nomenclature, composition, and function. From the revenue collections of the English collectorate, cash sums were disbursed to the Nizamat treasury; from there certain sums were in turn disbursed to each head of a de'orhi. The latter devolved into the broad groups of *mata'aynna* (perquisites of a post) and *ma'mulāt* (plural of ma'mūl or customary, fixed allowance) then paid to various male and female kin.

The system by which allowances and wages seem to have been disbursed varied between annual payments and monthly payments. Those servants who were paid by the month came under the category of *mahīnādārān*, sometimes also referred to as *naqadi*. Most payments were over a longer period, and were made up of allowances in kind, rather than cash. This meant that though listed as monthly allowances, they were paid in lump at the convenience of the dispenser of the stipends. Each stipend listed against a slave was also expressive of the amount expendable on the survival of the slave. In 1802, Diwan Manik Chand had explained to the Committee that in addition to the lump sums of money allocated to each de'orhi, another Rs. 10,047 was allocated yearly for the two hundred-odd chelas, slaves and others: the allowance of each was paid in cash on a calculation of the different articles and charges of apparel and their probable cost; that no individual received more than Rs 80 and none less than Rs 40 per annum.[7] As the numerous petitions of concubines and servants put it, there was a direct relation between the cash allowance of each and the slave's physical survival. One petition said, 'we drew a monthly sum of 30 Rupees on account of our victuals &ca. from the established allowances of the de'orhi of Her Highness the Babbu Begam deceased'.[8] Another slave-concubine from the household of the Nawabs of Dacca

[7] Statement Exhibiting Reductions in the Nizamat, submitted by the Committee under the Presidency of Mr Pattle, BC F/4/250/5602.

[8] Petition of Muhbooba, Fukrufroze and others in BPC, 12 July, 1836, no. 3 and 27 September 1836, no. 9.

claimed that 'a sum of 200 Rupees including a cash payment of 10 Rupees for Betlenut used to be expended on her account for all eatables, shoes, cloths'.[9]

Though money was a significant component of this larger realm of the gift, there were other components as well. Chief among these was cloth and food, two constants in the playing out of deference and dependence. Cloth both expressed the relationship between givers and recipients and differentiated among the latter by virtue of the distinctions in the qualities and amounts given to each. Perhaps because of the Company's own investments in textiles, its Agents were also more attentive to the amounts and kinds of cloth that was hoarded in, and disbursed by, each de'orhi. Thus, upon Manni Begam's death, an inventory of the property listed 'eight bales of cloths, 160 pairs of long shawls, more than 50 shawl handkerchiefs, 1200 pieces of muslin, 300 pieces of linen, Radhanagar and Cossimbazar silk goods and beautiful embroidered purdahs which cost her Rs. 20,000'.[10] The textiles were important because these were in turn given out under her patronage to the various dependants of her de'orhi. From the de'orhi of the Manni Begam alone, according to her Diwan, a sum of Rs20,807 for 'white cloths' annually (*sāliyāna*) was disbursed in the Mahals according to the degree of favour in which the individuals happened to be held at that moment. The principle underlying the access to different kinds of cloth in the Nizamat was relatively straightforward: the costliest was reserved for a small number of recipients and the cheaper, coarser varieties were for general disbursements. So the sahibzadas and *sahibzadis* (who traced descent from a Nazim) received greater amounts of the renowned and expensive Dacca muslins,[11] articles like the *bāluposh* (quilted covering) and *razāi* (coarser blankets) were never included in the allotments of

[9] Petition of Ushrufulnissa Hubos in Commr., Dacca, to Sec. to GOI, 9 August 1843, BC F/4/2091/97369.

[10] T.Brooke to Pers. Sec. to Govt. 12 April 1813, BPC, 6 July 1816, no. 46.

[11] For tests of fineness in muslins, the weight of the cloth in proportion to its length and the number of threads in each warp, and the place of each kind *Sharbat*, *Shabnam* and *Abrawan* in the range of muslins *see* T.N. Mukharji's *Art-Manufactures of India* (Calcutta, 1888, reprint Delhi, 1974), pp. 315–20 and *Dictionary of Economic Products of India* (Calcutta, 1890) *also* N.N. Banerjee, *Monograph on Cotton Fabrics of Bengal* (Calcutta, 1898).

the sahibzadas, but were regularly given to the khawas. Sumptuary distinctions in turn set the women of the *gaenpura* (singers and musicians), the *khawaspura* (personal attendants) and khwajaserai (eunuchs) apart from each other. The cheapest kind of razāi, the Azimabadi, went to the lowest-ranked women in the hierarchy of the mehal, whereas the slightly superior among the gaens in the Chougurrah received the middling-priced Radhanagri bālaposh and Bandri razāi along with single shawls. Furthermore, the single shawls of the gaens were of lesser value than the single shawls given to the concubines of previous Nazims. Only those who had borne sons to the Nazims received a *doshāla* (translated literally as a pair of shawls) valued at Rs 150.

The sensitivity of cloth as a barometer of stratification was not a reflection of its commercial value alone; it was a major insignia by which the wealth and standing of the *giver* was reckoned. Since to accept material goods was, at base, to acknowledge one's dependence on the giver, what slaves received—cash allowances as well as cloth and food—must be qualified as ways of motivating all of them to specific kinds of tasks and functions. It logically followed that the kind of clothes slaves wore, determined by various owners/masters, sometimes indicated the work they were meant to perform, particularly slave-attendants like the *sonthbardārs* (holders of silver standards) and the gāens. In both instances, rich or fine clothes (as well as jewellery) were worn on occasions when their owners wanted to advertise their wealth and taste.[12] In the political economy of deference, these clothes, food and allowances signalled their private dependence upon specific masters and the credit of the latter. Thus an astute observer in Awadh described the symbolic nature of the dressing of slaves:

> female slaves are often superbly dressed... richly adorned with the precious metals, in armlets, bangles, chains &c; the *lady thus adding to her own consequence by the display of her attendant slaves*. The same may be observed with regard to gentlemen, who have men-slaves attending them, and who are very frequently attired in costly dresses, expensive shawls and gold ornaments.[13]

[12] Other body decorations, like shaving of heads or boring of ears, appear to have been equally significant. Phrases like *halqa bagosh* used to denote the male slave with ears bored to carry rings, in the will of Soulut Jang, L/AG/34/29/42, I, p. 178 indicate this.

[13] Mrs. Meer Hassan Ali *Observations*, p. 41. Emphasis added.

In 1802, the Company's policies of retrenchment in the Nizamat were resisted on precisely such symbolic grounds by the Nazim. Referring to the annual savings proposed by the Nizamat Committee in the allowance for the chelas, slaves and attendants of the de'orhis, he showed that the drastic reductions imposed over the period of twenty years—from Rs 105,454 annually in the 1773 estimate, to Rs 57,653 in 1802—had led to the coarsening of the attire of slaves. Thus he urged that 'where the finest muslins *abruwan* and *shubnums* were before used to dress the household slaves,' latterly inferior types of cloth were used. As Nāsir-ul-mulk saw it, the dress or undress of the slave was a direct display of the credit of the master and the household:

> in the establishment of Chelas who became slaves on condition of food and raiment... every practicable reduction in their allowance has been made... should [further reductions] be effected, they would be left naked and without clothing which would be discreditable and humiliating to the sarkar.[14]

By 1824, the sartorial impoverishment of many of the slaves surrounding the Nazims was noted by other observers in the words 'ragged and tattered' used to describe the servants of the Nazim.[15] This indicated that the minute distinctions between different groups of slaves and servants were under threat. Indeed, even a young incumbent Nazim invoked the sartorial register to represent his own impoverishment : 'my wearing apparel, which is of the most common kind, and the lace etc. on it not silver but tinsel'.[16]

In large measure, the deteriorating material condition of the slaves and servants of the Nizamat can be traced directly to the policies and socio-legal ideologies of the Company's government in Bengal. Based on a broadly two pronged approach, separating 'kin' from 'servants', circumscribing the membership of the former while ossifying the composition of the latter, and privileging this circumscribed kingroup over the latter, the Company agents effectively diminished both groups over time. The agents' attitude to the stipends and allowances of concubines, slaves and servants in the constituent households of the

[14] Enclosure D., Nawab Nazim to President of Committee, 1802, BC F/4/250/5602.

[15] W. Loch AGG, to Stirling, 30 August 1824, *MNLI*, I, p. 313.

[16] From His Highness the Nawab Nazim, recd. 29 December 1827, BC F/4/1228/40156.

Nizamat was to commute 'gifts' of cloth and food into cash amounts, hive off this cash-allowance for a slave-servant (part of the mata'yanna of a relative) from the ma'mulāt, and treat each slave-servant's stipend as an 'ordinary life provision' rather than 'private property'. This meant that with the death of each, the money-allowance lapsed to the Deposit Fund (established 1818, and confirmed in 1834) regardless of any heirs he or she may have left. Hence, when some of the very old and long-serving slaves died, their stipends would not be renewed to any of their natural heirs, on the grounds that 'it was inexpedient that persons of this class should be encouraged to a life of idleness, and exempted from earning their own livelihoods'.[17] While this attitude to the slave-born sought to bypass the thorny issue of heirship to slaves and will be examined later, it was consolidated as policy by the 1830s. Thus when a slave-eunuch of Babbu Begam died, and the current mistress of the de'orhi recommended the division of his stipend of Rs 16 between two other slave-eunuchs equally, or when old female slaves receiving even smaller sums died, leaving both adoptive kin and biological heirs to whom the pittance was sought to be renewed in turn, the government came down heavily against 'making pensions to mere dependents hereditary'.[18] Unwilling to sanction the grant of any stipend where it could be avoided, the Government instructed the Agent to exclude particularly those 'dependants having no other connection with the family but that of service'.[19] Such pressures increased when the Company's Government tried to cut down on its disbursements from the Collectorate after 1818. When a daughter of a Nazim passed away, leaving many old slave-servants and dependants, it was not a matter of policy to grant pensions to these people: it was treated as an exceptional concession,[20] contrary to the opinions of

[17] Sec. to Govt. to AGG Martin, 28 April 1818, *MNLR* II, pp. 153–4.

[18] Sec. to GOB Macnaghten to AGG Cobbe, 5 March 1833, Extract Poll. Cons. of 5 March 1833, nos 59–63, BC F/4/1475/57976. For the disinherision of the adoptive son and brother of Bībī Fatehābādi, a dancing-girl and concubine of Nazim Nasir-ul-mulk, *see* AGG to Sec. GOB, 3 October 1836, BPC, 11 October 1836, no. 9.

[19] Sec. to GOB Macnaghten to Offg. AGG, 11 July 1833, Extract Poll. Cons. 11 July 1833, no. 147, ibid.

[20] AAGG to Sec. to Govt., 3 December 1833 and reply, NAI, FPP, 19 December 1833, nos 98–9.

earlier nazims. It was of such a group of 200-odd slaves (93 females and 100 odd male slaves and eunuchs) of the de'orhi of Babbu Begam, that Nazim Nasir-ul Mulk had said

> from a principle of compassion and consistently with justice, I can never consent to turn off those ancient servants to beg their bread from door to door.[21]

However, since English officers in the first half of the nineteenth century required that receipts of money be commensurate with services performed, and very little money be expended on slave-allowances generally, older slave-servants became vulnerable to Company-led financial manipulation. Such inconsistencies were supposed to have been wrinkled out by the terms of the compromise of 1834. Under the latter agreement, the Company confirmed its sole control over the lapsed stipends of previous and current gaddinashin begams in exchange for its nominal withdrawal from the appointment and dismissal of officers.[22] However, the Company's relentless pressure against the continuance of petty stipends, or the transfer of stipends from one slave-servant to another, eventually effected what previous Nazims had found unthinkable. In an economy drive under Nazim Humayunjah in the early 1830s, many of the very old dependants, including some chelas of an earlier Nazim, were dismissed, and having no other option, took to begging.[23]

A further deterioration of the living conditions of slaves in each de'orhi, and each household, occurred as a result of an increased fragmentation of stipends of those who counted as 'kin' in the Nizamat. The kin-stipends fluctuated wildly from time to time, because the Agents and the Nazims collaborated in stopping the payments of stipends to recalcitrant or non-deferential kinsmen and women.

[21] From His Highness, Mubarak-ud-Daula II to the Vice-President, recd. 15 December 1809, BPC, 30 January 1810, no. 41.

[22] AGG Thoresby to Dy. Sec., C.E. Trevelyan, 17 February 1834, BPC, 13 March 1834, no. 40; and enclosure for the clauses of the agreement of 15 February 1834. For Trevelyan's note of 11 March 1834, characterizing the resumption by Government of 6 lakhs—'by far the most valuable part of the Sixteen lakhs' as 'a great point gained', *see* ibid, no. 43.

[23] AGG Caulfield to Sec. to GOB, Prinsep, 12 October 1838, BPC, 17 October 1838, no. 7.

Furthermore, when such kinsmen and women died, and disputes began, the persons immediately affected were those slaves-servants-concubines who had the least power to manipulate the petty administration of the Nizamat and the Agency. Some slaves did leave on this account. When the Nazim Alijah died, one of the seven slave dancing-girls he had kept as his concubines fled from the de'orhi of Bahu Begam. Asked by the Magistrate and the Agent her reasons for doing so, she claimed that 'since his Highness's death she had suffered every sort of deprivation and had scarcely been supplied with the common necessaries of life'. When the Agent tried to reassure her that Rs 100 had been proposed for each concubine as a pension, this slave-concubine refused to return, declaring that 'nothing but force should induce her to return to the Mehal'.[24] If this could happen to a Nazim's own concubines, the situation was even more fragile for the slaves and concubines of the less-privileged brothers and uncles of the Nazim.

While assignments of revenues for the 'maintenance' of slaves was an integral feature of the household-polities of the late eighteenth century,[25] a further complication within the general system of allowances and stipends was their relation to different kinds of tenures and lands.[26] Characteristic of many slave-holding households in Bengal, the connections between slaves, different kinds of land-tenures, and the monetary allowances stipulated for each was a fairly convoluted one. In the Nizamat, the strategies of carving out private holdings separate from assignments of revenue attached to the office of the Subedar, for example, was aided by the registration of many such holdings (*benami*) in the names of slaves, slave-born women and men,

[24] AGG to Pers. Sec., 29 May 1823, BPC, 31 June 1823, no. 129.

[25] For sums of Rs. 193–17 annas allotted to the bhandaris (male slaves), attending upon the Zamindar of Nattore; sums of Rs 65–18 annas marked out for the bhandaris who attended the purohits ('Bramins') and Rs 15–1 annas allotted to the dasis of Rani Bhavani, *see* Report of the Committee of Circuit, Rungpore, 16 December 1772, unnumbered.

[26] For similar allocations of Rs 563–8–0 monthly for 150 bhandaris and Rs. 36 monthly for 18 dasis ('for beating wheat') in Dinajpur zamindari *see* Collr. Dinajpur to President, BOR, 11 July 1794, BOR Misc. (Wards), 25 July 1794, no.2; *also* Sirajul Islam, *The Permanent Settlement in Bengal: A Study of Its Operation 1790–1819* (Dacca, 1979), pp. 206–7, table 12, for 1152 bighas of *chākerān* lands, held by the (male) slaves of the Dinajpur Raj.

concubines and wives. The explanation offered for this peculiarity of Indian revenue-records, offered by the masters themselves, was that no Nazim, or respectable person, could engage in commerce without loss of honour and dignity. Yet others ventured as a reason for these 'fictitious Registries' the desire to 'protect the Nizamat family from the inconvenience and humiliation of attendance at the local Courts, which a Registration of Landed Properties in their own names, and the consequent responsibility of Proprietorship, might sometimes necessitate'.[27] Well into the nineteenth century, the Diwan Nizamat, Prosunno Narain Deb explained the purchase of promissory notes with the Nazim's money in the Diwan's name thus

> both he [the Nazim Faridūnjah] and his predecessors have thought it beneath their dignity to have their names set down in connection with any undertaking in which their names might be used publicly... landed property of any nature, the possession of which in his own name, might entail the use of his name in any of the public courts, are purchased by His Highness either benamee or in the name of eunuchs, or other confidential servants.[28]

Though writing the name of a dependent against a particular property or tenure (benami) was a widespread practice, in the instance of the Nazims it had added significance. Benami lent another nuance to the preserved sacredness of the Nazim by distancing him from the world of commerce; it protected *his* name from being sullied on the tongues of the vulgar and it left the scrutiny of his overlords to fall on those slaves, eunuchs and subordinates, whose task it was to shield and guard him. In situations of certain threat to the Nazim's or their mistresses' interests, these eunuch managers and agents tried to deflect danger by claiming the property as 'their own' or 'self-acquired' property. When certain estates in Midnapur which had been 'originally benami for His Highness the Nawab Nazim' were farmed to Lewis Tiery, of the Calcutta firm, Messrs. Watson and Company, by the AGG Torrens,[29] without due authorization, two eunuchs Nazar Ali

[27] D.J. Money, Judge, Diwani Adalat Murshidabad, to B.J. Colvin, Register of Sadr Diwani Adalat, 16 April 1852, *MNLR*, II, p. 721.

[28] Raja Prosunno Narain Deb to AGG Lt. Col. C. Mackenzie, 3 December 1861, WBSA, GPP, December 1861, no. 85.

[29] Seton Karr to Major Lang, Offg. Agent, 1 July 1851, *MNLI*, II, pp. 664–5. The portions of Midnapur which were supposed to have been held by the mortgagee

Khan and Amān Ali Khan procured the transfer of these lands to Nazar Ali Khan in order to eject Tiery. The eunuchs declared Tiery's sanad a forgery, and allied with the zamindar of Taki, Baikunthanath Rai Chaudhuri, promising him 'the Dewani of the Nizamut as the price of his assistance'.[30]

A particular kind of slave-agency thus made the relationship between slaves and inanimate or immoveable wealth vital to the material well-being of the Nazims and his kingroup. Thus according to a list prepared in the late nineteenth century, different Nazims, gaddinashin Begams, and heads of collateral branches (like Purneah, Chitpur and Dacca ruling houses) continued to acquire different mehals comprising both *lakhiraj* and *khiraj* lands and *patni* estates in the name of individual slaves.[31] Slave-eunuchs, as agents, acquired landed estates from others not directly of the Nizamat, as they had done in the late eighteenth century too. Thus Ambar, the eunuch of Babbu Begam, bought two revenue-paying taluqas (Chakla Batai and Parganah Patkabadi) in Nadia from the zamindar Raja Ramkrishna,[32] lands which continued to be held in the possession of the Begam. Similarly, a six-and-a-half-annas share of Dihi Daibargapara, belonging to one Batul Fatima Begam, was sold to the eunuch Jawahir Ali Khan, the nazir of Bahu Begam for 'a diamond ring and a copy of the *Al-Quran*' in 1243 B.S./1836. The remaining portion of the plot, the share of 2 annas and $7^1/_2$ gandahs., belonging to one Azīzunnisā Khanum was given by the latter to Jowahir Ali Khan in 'lieu of attorney-fees', for a case successfully conducted by him on her behalf.

proprietor, Nazar Ali Khan, contained 1522 villages, with an estimated area of 4,97, 053 bighas and provided an annual income of Rs. 123,988, *see* Chitta Panda *The Decline of the Bengal Zamindars: Midnapore 1870–1920* (Delhi, 1996), pp. 41–2.

[30] Torrens to Carnac, Magt. Murshidabad, 15 February 1852, *MNLR*, II, p. 687. From subsequent records it is evident that Tiery sued the eunuch for proprietory title to the lands in the Sunderbans: the eunuch Nazar Ali Khan meanwhile gave a *gantidari pottah* to Brojokishore Mitra for the same lands. By 1864, a suit to get possession had been instituted against the eunuch by the *gantidar, see Sutherland's Weekly Reporter*, II, pp. 168–70.

[31] Husain Ali Mirza, Nawab Bahadur of Murshidabad to Collr., 25 February. 1885, BPP, August 1885, nos 21–2, enclosed list nos 2 and 3.

[32] Paper delivered by Syed Comauludeen Hussein Khan, 7 January 1799, BRC, 1 March 1799, no. 40.

The onset of the resumption of lakhiraj tenures, part of the intensification of the revenue-generating drive of the East India Company after the Charter Act of 1833, revealed yet other lists of lakhiraj holdings in the names of slaves.[33]

Apart from the holdings on lakhiraj tenure, slaves also held other tenures on different kinds of immoveable property, like houses, garden enclosures, and ponds, within lands that theoretically belonged to the Nizamat. The important point to note about these tenancies was that each kind of land, residential (*bastu*) or non-residential (garden) paid a rent (*jam'a*) to the privy purse of the Nazim, albeit at a minimal rate. In other words, slaves were not just the cultivators of individually owned lands (*khas*), but themselves the tenants of such lands. This pattern of tenancies of khas mehal lands, by which lands were held by 'Relations and Dependants of the family at low Rents and Profits were appropriated to their own use' was apparently a widespread phenomenon.[34] In the khas mehal of Ramdaspur in Murshidabad, for example, there were many *raiyat* who were slaves and 'blood relations' of the Nawab Nazims, and the revenue administration of the Company hoped that the assessment of the 15,000-odd slave and dependent-kin raiyat would add Rs 6000 to the gross revenue collections of Ramdaspur.[35] Hence another twist in the declining political economy of the household was added when the Company's government in Bengal ordered the resumption (and liability to revenue-payments) of all such khas and lakhiraj lands. Yet extracting rents from these slave/dependent kin tenants was difficult for the administration of the Company. Even if the mehal was farmed out in *ijāra*, the prospective revenue-farmer offered a very low sum for it because of the difficulty of realizing rents from such raiyat.[36] So,

[33] 'Abstract of Lakheraj Lands', BPC, 1 November 1836, no. 5; Sec. GOB, Prinsep to Pemberton, 10 April 1839, *MNLR*, II, pp. 391–2, enclosure 2.

[34] Graham's Minute on the Report of the Collr. of Dinajpur, in BOR, Misc. (Wards), 7 July, 1795. A similar tenancy pattern for the khas lands of the Coorg royal household is studied in T.P. Vijaya, 'Honour in Chains: The Problem of *hitti-bitti-chakri* in *Jamma* Tenure in Coorg, 1800–1939', *IESHR*, 32, 2, 1995, pp. 135–53.

[35] Collr. Murshidabad to Sec. to Govt., Rev., 15 October 1836, BRC, 18 October 1836, no. 25.

[36] L.J.H. Grey to Commr. of Rev. for 14th Division, Berhampore, 8 October 1838, *MNLR*, II, p. 354, enclosure 3 in no. 983.

persuaded by the fact that each holding was 'trifling' and none of the holdings listed by the deputy collector were of the kind ordinarily understood to be cultivable lands to be let out to better tenants upon default in revenue-payments, the Government of Bengal sanctioned the exemption of lands 'held by blood relations' of the Nazims from the operation of the resumption laws by 1839.[37] Specifically left out of the privilege were those lands held by slaves like the eunuch Basant Ali Khan, slave-born sons of affinal males (dāmāds) like Mirza Askari, and indeed some of the dāmāds like Shamsher Jang himself.[38]

Very different issues surfaced together in the light of the Company's desire to resume (and tax) slave-held tenures and other wealth. The first was kinship, the privileging of relations of blood over relations of food and cloth. From 1802 onwards, the ambit of the aqriba had been gradually circumscribed. Only 'direct heirs' by lineal descent were to get stipends and allowances. By 1830, both 'collateral branches living separately' and co-resident slave branches were to be left out of the possibilities of negotiating their status on the basis of working the peculium.[39] The second was 'proprietorship'. Could slaves 'possess' wealth endowed them by masters? Some slaves were endowed with full proprietory capacities with respect to small portions of immoveable wealth. For the remainder, particularly wealth made up of promissory notes, lands, houses, and plate, only very limited capacities (of managing and enhancing, not alienating or disposing of it) were enjoyed by slaves. The difference between management and proprietorship by slaves was visible in a contest between two blood-brothers in Chittagong in 1796 over the division of hereditary lands. The younger brother claimed that certain villages acquired by his grandfather as a gift had in turn been entered as Turruf Nychund after Nychund 'an industrious and attentive slave' to whom the *management* of these villages had been made over.[40] These villages, along with the self-purchased slaves of the grandfather, had in turn

[37] Sec. GOB to Sec. Rev., 29 January 1839, BSBR, 20 February 1839, no. 108.

[38] Sec. GOB to Sec. SBR, 23 October 1838, BRC, 2 April 1839, no. 32.

[39] Sec. to Govt. Stirling to AGG Dale, 5 March 1830, *MNLR*, II, p. 209.

[40] Progs. of the Diwani Adalat of Zillah Chittagong before J.E. Colebrooke, in case of Ramdulloll vs. Ramkishore of 28 June 1783, in Proceedings of the Sadr Diwani Adalat, 27 April 1796, nos 13–14.

been gifted to one of the uncles of the plaintiff, and therefore had not been subject to further division among heirs. The elder brother's rejoinder was:

There was a slave of my Grandfather by name Punchaun who having acted as Khedmutgaur to the Lala and acquired some money by his industry purchased *at his own expense* villages Chorawtee Gyah &ca. which he caused to be entered at the Sudder as a Turruff and after paying his Revenue whatever profits remained, he appropriated to his own use. *He with his own money bought Four slaves Nychand Chandah &ca. for his own service....*[41]

The elder brother emphasized the fact that the purchased villages were 'only given to the slave the management of them that he may as nankar obtain something for his support.' In Murshidabad, while some agents and English barristers accepted the contingent nature of 'possession' by slaves,[42] other officers were wholly at sea with such ideas. Hence an official, who had served as assistant to the magistrate of Murshidabad, made two contradictory statements about 'slave-property' in the same breath. Slaves, Cheap wrote, purchased their freedom from their earnings,'thus establishing that the property possessed or realized by a slave in servitude is his own'. At the same time, the property of a slave reverted to the master upon the death of the slave, but slaves seldom had anything to leave of their own, 'their clothes and every thing they have on, or what they have for use, belong to their masters, who provide them with everything'.[43] The confusion about proprietorship by slaves added to the problems of inheritance from slaves.

Wealth and Its Aftermath

The Company's policies on 'slave-proprietorship' of allowances, tenures, promissory notes, houses vacillated according to a) the estimated size of the wealth in question, b) time c) gender and d) the determination and attitudes of specific agents, and came to a head

[41] Ibid., 30 September, Deposition no. 5. Emphasis added.

[42] Enclosures A–B, in AGG Melville to W.H. Macnaghten, Sec. to GOB, 20 January 1837, *MNLI*, II, pp. 55–9.

[43] Appendix II, Report from Indian Law Commissioners, (*PP*), 1841, no. 47, p. 293.

particularly around issues of inheritance from slaves. For benami and slave-agency to work smoothly and in favour of the real proprietors, the Nazims and the Begams needed to ensure that such wealth was not secretly siphoned off by the slave or dependent, who held it temporarily, and even if it was that it would almost inevitably be returned to the Nazims by virtue of the operation of inheritance norms.

The place of wealth in the transmission of dependence and negotiations of independence in the Nizamat was complicated by two different problems. One was the incompatibility of the Company's revenue-extraction measures with the maintenance of practices of slaves' temporary holdings and capacities. The second was the absence in English slave-law of a device shared between Mediterranean and Near Eastern slave-holding societies which allowed manumission and affiliation of the slave with the masters' kingroup to be exchanged for permanent loss of proprietory capacities. The device through which this incapacity was articulated in the former was the *paramone*, and in the latter it was the institution of *wala*. Strictly speaking, the paramone was the provision of support by a slave to the master/ mistress and acting as a condition for the former's manumission, a form of retaining the services and income from the slave but not responsibility for his/ her delicts.[44] This kind of paramonar manumission seemed to have been widely prevalent and may be compared to the relationship of wala between manumittor and slave in Islamic law.[45]

The main legal incidents of wala were (a) title to the freedman's estate, on which there was no disagreement between either Sunni or Shia law. The differences arose between the two schools in the ways they represent this title: in Sunni *fiqh*, the master inherited as the last agnate of the freedman, taking the *residue* after Quranic heirs, to be

[44] For an instance of this kind of manumission, *see* Charu Chandra Ray, 'Slaves and Slavery in old Chandernagore', *BPP*, 6, 1910, pp. 257–65. In one deed, Antoine Manhon, 6/7 year-old slave, 'native of Balasore', was freed on the promise that he would pay his owners/masters Rs. 6 per year for the period of six years, as repayment for the sum expended on him. The repayment was facilitated by his being 'bound over' to the merchants of the Royal Company of France.

[45] For the historical development of different kinds of wala, *see* Daniel Pipes, 'Mawlas: Freed Slaves and Converts in Early Islam' in John R. Willis (ed.) *Slaves and Slavery in Muslim Africa* (London, 1985), I, pp. 199–239 and Patricia Crone, *Roman, Provincial and Islamic Law: The Origins of the Islamic Patronate* (London and New York, 1987).

excluded by a biological agnate and himself exclude remoter relatives, like a sister's son. The Shias, however, held that the manumittors' title rested on *sabab* (tie, connection, derivation) rather than nasab (blood-relationship) and excluded the manumittor from succession in competition with any blood-relation of the slave/freedman, regardless of proximity of relationship. These differences would be particularly relevant to the claims of children born of female slaves, and their descendants. Other obligations of wala were (b) the obligation of paying blood-money on behalf of the freedman, and (c) the claims of manumittors to act as marriage guardians to the freedwoman or the freedman's daughter. The manumittor's rights and duties extended to the freedman's descendants, as well as to his freedmen and their freedmen in perpetuity. And on his own death they passed to his descendants in perpetuity. This relationship could not be the subject of sale, gift or other kinds of alienation. Thus at these points the wala relationship came closest to resembling a kinship-tie.

The difference was that the rights and duties vested in kinship ties were usually reciprocal whereas in the wala, it was unilateral. The manumittor acquired rights and duties *vis-a-vis* the freedman but the freedman did not acquire the same claims upon the manumittor. In other words, wala as an institution continued the relationship of dependence between slave, freedman or woman and master or mistress, through the maintenance of specific incapacities on the part of the 'lower *mawla*' (the ex-slave) and the retention of specific capacities in the 'higher mawla' (the manumittor). As far as it affected female slaves, the Quran, legal opinion of men learned in the sharia and household practice concurred in that the slave who had borne a child to her master became his *umm-i-walad*, the mother of his child, a status that ensured, in theory, her immunity to further sales and entitled her to emancipation at his death.[46] However, these privileges were differently accommodated in the different schools. The Sunnis considered these permanent but the Shias held that the exemption from sale was restricted to the life of the child a slave had borne to the

[46] *Encylopedia of Islam*, I, p. 31. For a study of the impact of such shared codes on the sales of pregnant slaves, *see* Ehud R. Toledano, 'Slave Dealers, Women, Pregnancy, and Abortion: The Story of a Circassian Slave-Girl in Mid-Nineteenth Century Cairo', *Slavery and Abolition*, 2, 1, 1981, pp. 53–68.

master, and her title to emancipation was at the expense of her child's share in the master's estate.

WALA AND THE SCHOLAR-OFFICIALS OF THE COMPANY

Though specifically associated with the manumission of slaves, and granted to the manumittor as a compensation for his/her 'gift' of various capacities, wala, expressing the master's ultimate control over the slave's earnings, was retained by the different schools of Sunni and Shia fiqh. Given this unanimity in doctrinal law, the silences and suppressions of Company-found law are important. In the translations of the *Sirajiyyah* and of the *Hidaya*, the discussions of walà had been relatively extensive but these discussions became fainter in the work of the official jurists of the Company in the first half of the nineteenth century. As a term, it is absent both from the glossaries appended to, and the body of the texts and commentaries of William Hay Macnaghten's *Principles and Precedents of Moohummedan Law*, a text published in 1825 and very widely used by the English judges of the law courts of the Company.

The absence of the term wala from this influential text has to be explained before we can proceed to see its consequences for the Nizamat and such slave-holding households. One of the possible reasons for this may have been semantic, for as eighteenth-century English lexicographers noted,'There is no single word in our language fully expressive of this term. The shortest definition of it is, the *relationship between the master* (or patron) and his *freedman*; but even this does not express the whole meaning'.[47] Furthermore, the silence may have been the consequence of the term's absence from pleading itself so that if a claimant did not use the term it may not have entered the early English jurist-official's field of enquiry. Cheap, Judge of Mymensingh was categorical that claims to the property of deceased slaves had never become a subject of inquiry in a criminal court, and he had no instance of this from the zillah court he presided over.[48]

[47] Francis Gladwin, *A Dictionary of Mohammedan Law and of Bengal Revenue Terms with a Vocabulary, Persian and Vocabulary, Persian and English*, (Calcutta, 1797), p. 52, no. 755, 'willa'.

[48] Appendix II, Report of the Indian Law Commissioners, (*PP*), 1841, p. 293.

On the other hand, a close look at some of the precedents cited by Macnaghten seem to suggest that, in practice, wala was claimed, even though the term is absent, and the reply of the Islamic law officers to the cases seem to suggest that they recognized the claims for what they were. In one dispute, the Qazi cited a quotation from the *Sirajiyyah* which stated in full the degrees of proximity by which inheritance is meant to be distributed, including the right of 'the master of an enfranchised slave'.[49] In the section of Principles of Inheritance in the same book, the discussion of the distant 'kindred' specified that if 'the estate to be inherited belonged to an enfranchised slave, his manumittor and the heirs of such manumittor inherit, in preference to the distant kindred of the deceased'.[50] This is the closest Macnaghten got to the institution without its name.

A further reason for a general non-appearance of the term may also have been the result of procedure. In the majority of 'civil' cases submitted by the provincial courts to the appellate jurisdiction of the Sadr Diwani Adalat, the statement of the case by the lower-level judge would already have framed the issue for judgement. This is where the local official's definitions of who was or was not a slave would impinge on the submission; all too often, the reference to the Qazis and Muftis obliterated the fundamental relationships between persons, referring to them as A,B,C ad infinitum, *as though* each individual was jurally 'free' and socially autonomous. In the process, the wide range of capacities and claims that corresponded with gradations in slave-status were erased from the British judicial and legal records. Any remaining chances of inferring capacities, on the basis of relationships between persons, was attenuated further when the substance of slave-status was depicted in the language of kinship.

The likeliest explanation for the suppression of wala in Company-established courts was political. In Shia law, the failure of heirs meant that the property in question fell to the share of the Imam. The only equivalent for this in English law was the term 'escheat'; it was a term liberally used by the early Company men in their discussion of land-tenures, and in deciding whether the land belonged to the sovereign,

[49] Macnaghten, *Principles and Precedents*, Case XXVII, pp. 103–5.
[50] Ibid., p. 8.

in this case, the Mughal Emperor.[51] After 1765, having received the Diwani from the Emperor, it became even more critical to many functionaries of the Company to acquire those attributes that characterized, in their eyes, Mughal sovereignty. Claiming the rights of the Imam tallied almost too neatly with the revenue interests of the Company after 1773, when, with the passing of the Regulating Act, the control over its revenues became subject to the governance of Parliament. In the range of measures taken to ensure the stability and volume of revenues from India, a process just as important as the reification of land, occurred in the case of the Nizamat—the obliteration of boundaries between 'private' property and *Subahdari jagirs* (state property). The broad interpretation that the Company put on the treaties of 1765–66 was that the Nazim Mīr Ja'far had given up all his jagirs to the Company in return for a fixed monetary stipend. There was little formal, or treaty-based, recognition of the custom by which all Mughal officials and *subahdars* had tried to create private holdings separate from those which were attached to their office. As one scion of the house explained it in the late nineteenth century, the officials 'were obliged to do this, as otherwise, in the event of any withdrawal of the favour of the Emperor,... they would have been left penniless'.[52] The only lands the Supreme Council in Calcutta informally recognized as the 'private' holdings of the Nazim were the *ramnah* (lands attached to hunting lodges, and including grazing lands). All other kinds of wealth, jewellery, lands, furnishings, houses acquired either by direct purchase or inheritance or gift, either by the individual Nazims, or in their private capacities by members of the household, were deemed by all Agents and Company officials alike to be State property, that is, property held by the government of the Company. While explaining the avidity with which the resources of the Nizamat were scrutinized, this obscured a more fundamental haziness about the contours of the Nizamat, and about the relationship between the Nazims and the

[51] M. Athar Ali, *The Mughal Nobility*, pp. 63–6, argues that the operation of escheat was extended under the Mughals from slave-born to free-born members of the administration; *also see* Richards, 'Norms of Comportment Among Imperial Mughal officers' in *Power, Administration and Finance*, p. 264.

[52] Husain Ali Mirza to Collr. of Murshidabad, 25 February 1885, BPC, August 1885, nos 21–2.

Company. Was the term 'Nizamat' to express the unit of the Nazims and their relatives (aqriba) or that of the Nazims and the Company? Were the relations between the latter to be that of equals, or was the Nazim a vanquished Mughal official, to be subjected to the same norms and processes as any other subject of the British Government? In the initial years after the assumption of the Diwani, the Company operated within an ideational edifice according to which the Nazims of Bengal represented one (judicial) aspect of the State and the Company another (revenue). After the establishment of courts under Company control from 1790, this edifice too crumbled. Since the maintenance of law was no longer the jurisdiction of the Nazims, the relationship between the Company and the Nizamat had to be put on a new footing. Section X of Regulation XVI of 1793 sought to ensure some minimal powers of arbitration for the Nazim in his capacity as criminal judge, and implied that he shared some of the attributes of sovereignty. This in turn set up a grave political and philosophical issue for the Company—little realized at the time. Was the Nazim then a sovereign prince, whose person lay beyond the jurisdiction of the Company's law-courts or a 'private' person to whom the decisions of the English courts were applicable? The varying answers given to this in turn complicated the guarantee held out by the Company in 1773 that it would uphold Islamic law in matters domestic—marriage, inheritance and succession.

Company functionaries took fluctuating stands on all these issues at different times. In 1810, when an incumbent Nazim fell out with his brothers over the inheritance of the stipend-wealth and property of the deceased Babbu Begam (a grandmother in common to all the male claimants), the Government of Bengal upheld the right of the Nazim to appropriate the entire property by posing this as a political right. So it was said

> in the plenitude of the Nazim's power... would he have admitted of the distribution of his female ancestor's property among the heirs according to the Laws of Inheritance or would he according to the practice of Asiatick princes have considered it as revertible to the coffers of the sovereign?... there can be little or no doubt of the latter.[53]

[53] Pers. Sec. to Suptdt. of Nizamat Affairs, 20 April 1811, BC F/4/372/5257.

By 1813, this changed. A claim by Nazim Alijah to inherit the wealth of the deceased Manni Begam was resisted by the Company. The Governor-General's claim that the stipend of the ex-slave, Manni Begam, lapsed to the Company, rather than to the heirs and descendants of her manumittor i.e. the brothers, sons and grandsons of Mīr Ja'far, clearly overlooked 'Islamic' law. The Company's Government built up the Deposit Fund from 1818 by resuming the Begam's stipend, rather than allow the Nazim, or elder collateral males, or successive gaddinashins to inherit it, even though in terms of Islamic law, they were the 'legal heirs'. By 1820, when the slave-born daughter of a previous Nazim died and the incumbent Nazim, Walajah, claimed 'the right... to appropriate to his own use the property of the Dependants of the Nizamut', the Council in Calcutta refused to sanction what they called a 'pretension' on the Nazim's part. They argued that 'all such property ought properly to be regarded as belonging to the Nizamut and applicable... for general benefit of that Estate'.[54] While this aspect of 'escheat' law was reserved for the Company and denied to the Nazims in the first part of the nineteenth century, Regulation XIX of 1825 sought to protect the Nazim from the direct authority of the English-controlled courts. Justified on the grounds that it is 'not proper that His Highness should sue and be sued as an Ordinary Individual' this Regulation provided that no subject of the Company could bring a suit against the Nazims directly, and that judicial processes in such suits would be served on the AGG, who would conduct such suits under the authority of the Governor-General in Council.[55] This acknowledgement of his political persona could be equally dangerous for the Nazim. If, as some officers and agents held, the Nazim was not an ordinary, private person, it implied that unlike private persons who had personal wealth subject to inheritance laws, the Nazims did not have anything other than 'public property'. In other words, no Nazim could be allowed to sell, gift, or lease any of this 'public property' as an individual,[56] and government

[54] Pers. Sec. to AAGG, 23 October 1820, *MNLR*, II, p. 198.

[55] For text and discussion of the Regulation, *see* AGG Melville to Pers. Sec., 28 April 1825, BPC, 30 September 1825, nos 97–101.

[56] This was the position articulated by the Advocate-General, cf. AGG to Solicitor to Govt., 4 June 1872, NAI, FPP, October 1872, nos 266–7.

was not called upon 'to admit of Mahummadan Law being introduced into a Political Establishment over the affairs of which it has hitherto never been allowed to exercise any influence'.[57] The Nazims were thus gradually barred from claiming any share of inheritance according to Islamic law, (and would be led increasingly to establish their claims on the basis of a political authority which was more symbolic than substantive) as much as they were obstructed from the commercialized exploitation of the lakhiraj lands, or houses, they held.[58]

In addition the Company's ability to implement its guarantees regarding the maintenance of Islamic law was questioned whenever stipends and inheritances had to be divided between different claimants in the Nizamat. Was an Islamic law of inheritance to provide the guidelines of decisions? If so, which? Most members of the households receiving stipends tried to plead according to precedents within the Nizamat to find that precedents changed at will; the Government of the Company increasingly moved towards greater and greater suppleness in adjudicating fiscal claims. Thus in 1811, it would admit that the division of a stipend after the death of a stipendiary fluctuated 'either according to the Mussulman Law of Inheritance or agreeably to the principle of equity. The latter has generally, though not uniformly guided the distribution of the hereditary pensions.'[59] However, even principles of equity could be compromised for a disciplinarian end. The inheritance of a stipend by 1829 was to be made according to 'claims founded upon personal merit and good conduct, or the reverse, as some check upon the slothful, profligate and degenerate habits... among Nizamut stipendiaries'.[60]

A related complication arose since the Nazims were followers of I<u>th</u>na Ashari (followers of the twelve Imams) Shi'ism, at the same time that canonical, public prayers were conducted according to the Sunni traditions associated with the acknowledgement of the supremacy of the Mughal Emperor. In the early years after the

[57] AGG Caulfield to Prinsep, Sec. Poll., 18 December 1838, BPC, 2 January 1839, no. 4.

[58] Junior Sec. GOB to Sec. GOI, 7 May 1872, and Petitions of Lutchmeeput Singh, 7 June 1872, NAI, FPP, October 1872, nos 279–316.

[59] Pers. Sec. to Suptdt. of Nizamat Affairs, 27 September 1811, BC F/4/395/10033.

[60] Poll. Letter from Bengal, 10 April 1829, BC F/4/1186/30771.

establishment of the Diwani, the Company's servants did not need to distinguish between the different schools of Islamic law applicable to the members of the Nizamat. Manni Begam's countenance of a 'change in the Established forms of Mahomedan worship' in 1810-11 caused a certain group of Sunni inhabitants of Murshidabad to protest about the 'alteration of the form of prayer' sanctioned by the 'King of Delhi, which has always prevailed in Hindoostan, and which in fact he is bound as Nazim to maintain'.[61] Notwithstanding the general willingness of the Company officialdom to follow Manni Begam's directions in most matters pertaining to the Nizamat, the Governor-General intimated to her the intention of upholding Sunni law since it was the one 'by which the decisions of the Courts of Justice when under the superintendence of the Nabobs of Bengal and since that time, under the authority of the British Government have been regulated in all causes'.[62] Yet the Begam's own response showed little theological dogmatism. Accordingly, where she had required the Sunnis in Murshidabad to pray 'under the fly of a tent' at the periphery of the Nizamat *masjid*, she 'waived her objection to the ceremony being performed in the Musjid' on being handed the letter from the Governor General.[63] The apparent divide between personal faith and leadership of public prayers, as well as the easy negotiation between both Sunni and Shia forms lay at the heart of the problem. For apart from the syncretic *Behra* festival, when illuminated paper boats were floated on the Bhagirathi in homage to Khwaja Khizr to guard the waterways and protect travellers, there were many indications that underneath a formal Shi'ism, many life-cycle rituals were non-classical. For instance, alongside references to circumcision ceremonies of young boys were references to horoscopes prepared by 'Hindu' astrologers at the birth of a child.[64] The birth of a child, especially to a slave-concubine favourite of the ruling male, was prefaced by the *haftmasih* and *nehmasih*—ritual offerings of sweetmeats among all relations and friends in the seventh and ninth months of pregnancy, common to all

[61] G.Swinton, Dy. Pers. Sec. to Govt., to R. Rocke, Suptd. of Nizamat Affairs, 18 December 1810, *MNLI*, I, pp. 37–8.

[62] Enclosed in Swinton to Rocke, 1 January 1811, ibid.

[63] R. Rocke to Pers. Sec., 3 January 1811, BPC, 19 January 1811, no. 97.

[64] Letter of Mobaruck Ullee to AGG, n.d., BC F/4/1228/40155.

communities in Bengal.[65] Similarly, among funerary rites of deceased Nazims were gifts of elephants and gold *mohurs* to Brahmins[66] ; along with the *Shab-e-barat* and *Id* expenses were listed those of *Holi* and *Diwali*, and regular donations to the Vaishnava *akharas* at Sadiqbagh in Murshidabad. In a household-polity where Shia, Sunni and syncretist elements coexisted, by what school of law were conflicts to be decided? Given the fluidity and complexity of faith and ritual practices in the Nizamat, the lack of correspondence between the faith/customs of individuals of the household and the Sunni law upheld by the law-courts established by the Company formed a real obstacle in the Company's fulfillment of the commitment of 1773 to administer issues of inheritance and succession according to Islamic law.

These convolutions conditioned the selective erasure of wala and the reconstitution of 'Islamic' law in the Nizamat. The fact that heirs of masters themselves fell out over the inheritance from slaves and freedmen made it far more likely that solutions would be found by the Company's agents in terms of the secular exigencies of the moment rather than in terms of doctrinal law. In the Nizamat too, 'intra-family' disputes over shares of 'inheritance' occurred particularly in the division of stipends, as in the the conflict between the Nazim in 1810 and his uncles over the wealth of Babbu Begam,[67] and repeatedly over eunuch-held wealth. Most visibly in the case of the eunuchs, and male slaves, 'gifts' of jewels, cash or lands made by a previous master or mistress were disgorged by subsequent masters who inherited control over the same slaves. Thus the male khawas who had been given gold mohurs and precious stone as 'rewards for past service' by one Nazim were compelled to deliver up the gifts by his successor.[68] Later a Nazim claimed the estate from which lands had been gifted by Babbu Begam to her eunuch Roz Afzūn Khan in opposition to the Agents who

[65] *See* Evidence of Zamurrud Ali Khan, eunuch Nazir of Amīrunnisa Begam, 2 May 1836, and of Dārāb Ali Khan, Eunuch Nazir of Nazim Humayunjah, 4 May 1836, IPP, 1 August 1836, no. 84.

[66] W. Loch to A.Stirling, 1 November 1824, *MNLI*, I, pp. 551–2.

[67] Welland to Brooke, 21 February 1812, *MNLR*, II, p. 74.

[68] J. Ahmutty Commr. Murshiddabad to Offg. Pers. Sec., 17 September 1822, BC F/4/851/22643.

sought to sell them.[69] The theoretical basis that such a Nazim urged in extenuation of his practice was the same that caused another female lineage-member to resist her brother's slave-concubine. When the son of a Nazim died in 1827, his khawas-born daughter was granted the sum of money out of which the other slaves and dependants of the household were to be 'maintained'. When the daughter too died, and her khawas-mother Harunnisa' claimed the sum, she was resisted by the sister of the dead man on the grounds that Harun was 'A slave Girl of my mother... in the service of my late brother'.[70] To this, Harun responded that she had been married to the father of the child. Witnesses for both contenders were presented; the incumbent Nazim supported the heir of the masters by decreeing that a 'haram' could have no claim.[71]

In both the eunuch's case, and in the case of the slave-mother Harun, the principle issue was heirship from slaves and ex-slave. Who was the heir of each? In the case of the female, the dilemma was enhanced by the fact of her motherhood. Emancipated by the bearing of children, could she have been endowed with the ability to direct and use this wealth independently of erstwhile masters, and current affines? Could she have been heir to her own daughter? Though instances like Harun's were very often framed by English judges and officials as issues in the legitimacy of the slave-born, (and often tried by English judges as disputes about marriage) and the heirship of the slave-born to 'ancestorial' wealth, the epicentre of such conflicts lay elsewhere : in the efforts of heirs of masters to benefit from heirs of slaves, and to control their lives. In sum, such disputes were essentially about that little-named doctrine of wala, about social heirship from slaves, which vitiated the relationships between a) the Nazims and the Government of the Company b) members of the same household upon the death of a master or erstwhile slave and c) the members of the collateral households and families with those of the Nazims, the colonial authorities and the Begams. This particular doctrine was

[69] AGG Cobbe to Nawab Nazim, 16 June 1835 and 7 September 1835, and AGG Macleod to Nawab Nazim, 9 March 1836 and 12 March 1836, *MNLI*, II, pp. 9, 14, 23.

[70] Memorial of Jigri Begam to Lord Auckland, 12 March 1836, *MNLR*, II, p. 269.

[71] AGG Cobbe to Sec. GOB, 18 January 1836, BPC, 2 February 1836, no. 7.

elucidated by Nazim Humayunjah with regard to the estate of a deceased eunuch thus:

it has been the custom that the Property of the Chelas should go to the nazim... At the time that Bussunt Ulee Khan was bought as a slave, he was poor and had no property but Numma Manee Begum having made him Nazir over her household and having given him all the Property of Yatbar Ulee Khan Khoja who was formerly Nazir, he became possessed of a great deal of money during the time he belonged to the household and was the slave of Nummah Munee Begum.[72]

By this reckoning, the wealth of the slave was a *temporary* endowment by a mistress, (herself umm-i-walad), of the capacity to improve and manage resources on the condition that such a capacity ended naturally with the physical death of the slave. Since capacities were independent of pathological transmission, the material wealth managed or held by an eunuch or other slave was transferred according to the wishes of a master or mistress. Thus one slave after another could be endowed with the same capacity, through the transfer of the office or salary, or the holding concerned. This mode of transferring however, meant that ultimate direction, and the final fruits of such grants, came to the masters, and hence passed to their heirs. The disputes over the deceased slave's or freedwoman's inheritance were acrimonious because it involved the prior questions not of slavery or freedom, but *whose* slave the deceased had been. However, the forms that the disputes took differed according to gender, as only female slaves retained the capacity of becoming genealogical kin, and acquired a degree of emancipation as a result. Hence, more often than not, the claims to emancipation in the case of female slaves were entwined with the rights of their slave-born children to inherit a share in the estate of former masters / husbands / fathers of the children. Such claims in turn were resisted by other heirs of the same deceased males, whether free-born widows, or siblings. Hence, while the genealogical heirlessness of slave-eunuchs was obvious—and social heirship from them was disputed *among* the free heirs of a master—where female slaves were concerned, the disputes occurred between those born of

[72] From the Nawab Nazim to Lord Bentinck, 29 January 1834 in Extract Poll. Cons. 6 February 1834, no. 106 in BC F/4/1522/60090A.

her and other heirs of free masters. The matter was complicated further by differences in the locations and destinies according to a), the fact that some female slaves had children with other slave-servants, and had biological heirs who were also servants b), the fact that some female slaves bore children to the Nazims, their uncles and brothers, and had biological heirs who were free, with both vertical and horizontal claims of kinship within the Nizamat and c), female slaves who reared children in mitigation of their own 'heirlessness' and had 'adoptive' heirs d), those who had borne no children at all. Contests over inheritance in two (a and b) out of the four variables above involved discrete but inter-twined strategies. For the heirs of masters, it required establishing descent from, and heirship to, erstwhile masters and hence inheriting titles and claims over specific slaves while denying the claims of slave-born children to paternity of free males, in order to prove the failure of the slave-born's title to 'inherit' a share. This meant that unless the slave-born members were in turn supported by some one faction within the household, their chances of being acknowledged as kin were fairly remote, and the ability of their slave-mothers to become autonomous of their 'in-laws'/ masters negligible. A glimpse of some of the disputes in the first half of the nineteenth century will illustrate these issues.

Disputed Status, Convoluted Resolutions

The theoretical and material possibilities that opened up to female slaves who became mothers of children to their masters are important to touch upon before we can see the infarctions and infringements involved in the case of slave-holding households like that of the Nazims in early nineteenth century Bengal. Though framed as issues of Islamic law of marriage, the underlying claim was that of the slave-born son or daughter to be declared an heir to a free father. If Sunni, the slave-born's claim to be an heir to a deceased male was important to their control of 'social capital'—a simultaneous assertion of belonging in the lineage of their fathers, of bilateral genealogies and depths of ancestry, of the ability to direct other human beings through the direction of inanimate wealth and so on. If such a household was of the Shia persuasion, a declaration of the 'share' of the slave-born's inheritance could also be tantamount to the price of his/her mother's freedom, that is the amount of his share, expressed in monetary terms, could be traded in for jural autonomy. In a suit begun in February 1796, slave-mother Zeinub Beebee on behalf of her minor son, Himmut Ali, brought a claim for a half-share of the 5-annas, 12-gandas zamindari of the late Jafur Ali in the parganah Seriaul, Mymensingh.[73] She sued for her son's right to share equally and jointly with Hasan Ali, another son by a different mother, Zebonissa Begam.

The maulvis and qazis, referred to by the Sadr Diwani Adalat, gave an intricate and elaborate computation of shares, based on differences made by the order of death and the relationship of each survivor and distinguishing between claims of dower and claims of inheritance from husbands. Such answers proved simultaneously too complicated, and prolix enough, for English judges. Fashioning a simplistic, and superficially liberal, judgement which said nothing about whether this was a Sunni or a Shia household, they found a) that the son of Zeinub the slave was entitled to inherit as other sons, *but* b) as the share of the dower which was demandable from the estate and to which Hasan Ali was heir, absorbed the whole of it, and claims to dower had to be satisfied before partition of heritage, the slave-born son's share of the estate by inheritance *from his father* could not be admitted. Hasan Ali was 'enjoined to afford a maintenance to Himmut from the estate.'

If the slave-woman had been accepted as wife by the English judges, an option that the qazis' answers gave them, she would have inherited a share from her husband and previous master, and in theory, at least, passed on her share to her son. The English judges of the Sadr court upheld the slave-status and non-wifehood of Zeinub, circumscribing the potential of the slave-born son to assert autonomy of the authority of the freewoman's son. For slave-mothers the theoretical basis for bringing claims of their sons to inheritance were critical, since the child born of her free master and herself would not only inherit freedom of status, but also gain opportunities for negotiating her freedom from the continuing dependence upon the heirs of the free master. Therefore in selectively choosing between different schools of Islamic law, English judges and officials in the first half of the nineteenth century may have impeded processes of emancipation and

[73] S.D.A., I, pp. 48–52.

incorporation of the slave-born at one extreme, while leaving the larger panoply of masters' authority over slaves intact at another.

Claims, and denials, in court of marriage and inheritance claims of slave-wives and slave-born children almost always occurred as part of ongoing strategies from within specific authority structures and disputes within the household. As seen in Chapter 3, specific elders and superiors within the harem gave slave-born daughters in marriage to specific men chosen by them. Where female slaves were concerned, it mattered greatly whether they were first-generation slaves, or mothers of 'daughter-like', 'son-lịke' assimiles. Claims of dower or inheritance made on behalf of one woman or another had invisible beneficiaries, who were themselves rarely brought to court. Such claims in the Nizamat conditioned the conflict between dāmāds and Nazims on one hand, and the de'orhi elders and the males on the other. In October 1808, the Superintendent of Nizamat Affairs reported the death of one of the dāmāds, Rustam Jang[74] and conveyed his suspicion that the 'considerable property' of the deceased, claimed both by the reigning gaddinashin, Manni Begam, and by the Nazim, Nasir-ul-mulk Babar Jang, would be better applied to 'afford a comfortable provision' for the slave-born children who would otherwise become 'burthensome' on the Company. The Government of Bengal directed the Superintendent to resume the dāmādi allowance of the dead man. As for the rest of the property, the claim of the Nazim was based on the fact that Rustam Jang's wealth was a result of his marriage with his father Mubarak-ud-daula's sister's daughter. The basis on which Manni Begam claimed the property of the dāmād Rustam Jang, when the blood-related heirs of the 'father' of Shahrukh Begam were alive—like Khalilullah's sons, as well as Miran's heirs like Saiyid Murteza—was the 'deed of Gift' by which the owner/mistress, Saliha Begam, transferred her 'establishment' to Manni Begam.[75]

Both claims were referred to the Qazi ul Quzat of the Sadr Diwani Adalat. As the Government saw it, recognition of the Nazim as 'legal heir' was dependent on the size of the wealth/ estate estimated to be

[74] Superinterdent of Nizamut Affairs to Sec. to Govt., 19 October 1808, BPC, 14 November 1808, nos 35–8.

[75] From Fuzeelutunnisa Begam, recd. 31 August 1797, BPC, 6 October 1797, no. 27.

held by the deceased person. In this instance, the estate was apprized to be substantial:

> it would have been right that property so considerable as 55,000 Rupees should be brought to account and applied to the payment of His Highness's debts to the Company. In no case... could His Highness be allowed to succeed to property of such extent without rendering an account of it to Government which... is entitled to claim the application of any available assets to the reimbursements of the amount [debt to the Company].[76]

This in turn led to considerable convolutions in the representation of doctrinal law; and the upholding of one over another school of inheritance. The Qazi ul Quzat and the maulvis gave a decision according to Sunni law, while observing that there was no objection to the recognition of Shia law should the two parties (Manni Begam on behalf of Shahrukh the concubine's daughter, and the Nazim on behalf of Shamsuddi, his father's niece) claiming the property wish it so. The Government of Bengal not only disallowed this accommodation in principle, it deliberately misinterpreted the decision of the Qazi ul Quzat. Thus though the *fatwa* of the Qazi decreed that all property proved to have belonged to Shamsuddi Begam either as marriage portion from her family or gift by her husband was to devolve on her husband and on her own relations on 'the male side', the Government of Bengal decided that the Nazim could not claim because he was not kin 'by the male side'.[77] This was a travesty of the truth: both the Nazim and Babbu Begam had claims on Rustam Jang because of their relationship to Mubarak-ud-daula. (Note that there were no claimants on behalf of Shamsuddi's father). Intent on the admission of Rustam Jang's own slave born children to the inheritance in order to preempt any claims on the stipend of the father on their behalf, the interpretation of legal doctrine itself was subordinated to secular exigencies. When the Superintendent reported that in the opinion of the Sunni maulvis of the Provincial Court for Murshidabad Division, the concubines taken in mut'a (Soortellee and Daodee) and their children were not entitled to any share of the dāmād's property,[78]

[76] Extract Poll. Letter from Bengal, 18 October 1809, BC F/4/304/7006.

[77] Translation of the Answer by the Muhammadan Law Officers of the S.D.A. and Monckton to Pattle, 28 April 1809, BPC, 6 May 1809, nos 87–8.

[78] Translation of Answer returned by the Law Officers of the Provincial Court of Murshidabad in Suptd. of Nizamat Affairs to Pers. Sec., 7 July 1809, BPC, 15 July, nos 65–6.

both the Government of Bengal and the Superintendent quickly changed their preference. Now the Shia law of inheritance was forwarded as the law by which matters were to be decided, because belatedly, it was discovered, that school allowed the wealth left by the deceased to pay for the maintenance of the slave-and concubine-born children.

In spite of holding the Nizamat hostage to Sunni canonical forms of public worship, the Government of Bengal proceeded to uphold a Shia law of inheritance, and apportion the estate of the deceased dāmād thus: 2/16th to the woman supported by Manni Begam as the sole surviving 'widow', Shahrukh, and 14/16th of the estate to be divided up into seven shares which included the children born of the '*mamtood*' women.[79] Fully aware that the interests of Shahrukh (and Manni Begam) were not compatible with the partibility of the estate in the manner proposed by the Government, the latter then invoked another doctrine from canonical text—the doctrine of *iqrar* or acknowledgement. Arguing that Rustam Jang always 'acknowledged his illegitimate offspring to be his children,... and as it was not to be supposed that he would have publicly pronounced them to be illegitimate',the Governor-General in Council decided that they might be considered to be entitled *even according to the Sunni law* to participate in their father's property,—'in the same manner as if they had been born in wedlock'.[80] Such cynicism affected the lives of both the slave-born daughter whose person and property was controlled by Manni Begam, as well as the other slave-concubines and their children. The Government knew full well that the claims of the slave-concubines taken in fixed-time contracts to a share in the property of Rustam Jang depended upon their ability to prove that the sums promised in payment had not been met. The Government of Bengal believed that it was 'probable... that they have no such claims since the sums given on those occasions being generally inconsiderable are usually paid at the time when the engagements are contracted'.[81]

Within ten days of this decision of the Government, a claim for

[79] Ibid.

[80] Monckton to T. Pattle, 3 August 1809, BPC, 5 August 1809, no. 91. Emphasis added.

[81] Ibid.

the payment of Rs 68,000 as 'marriage' settlement *was* submitted, on behalf of Shahrukh Begam, by the vakil of Manni Begam, complete with a *mahrnama* bearing the seals of other dāmāds,[82] upholding the payment of *mahr-i-maujjul* (dower payable on demand). This effectively stymied the attempt of the Company to arrange for a partition of the estate. This claim of 'dower settlement', referred to their Law Officers again, elicited the opinion that the claim for payment of the widow's dower superseded all rights of inheritance. It must be satisfied *before* any division of the property can be made amongst other heirs. Their hands forced thus, the Government reluctantly sanctioned Rs 200 out of the dāmādi stipend, for the maintenance of the concubines and their children. It would however bend no further. In 1810, when Manni Begam tried to channel the resumed Rs 1000 of the stipend in the names of the 'domestic dependants of the family of late Nawab Rustam Jang', the Government decreed that Shahrukh Begam had been given possession of 'ample resources to provide for the maintenance of her late husband's dependants'.[83]

An identical trajectory was witnessed in the case of another dāmād. In June 1809, the dāmād Nazar Ali Khan Shahamat Jang, the husband of Rahīmunnisā, died, leaving four sons, five daughters, and six 'wives' apart from Rahīmunnisā .[84] In the conviction that the wealth left by the deceased could be divided on the basis of the kind of 'compact'

[82] Translated letter from Manni Begam to Pattle, recd. 17 August 1809, in Pattle to Swinton, 25 August 1809, BPC, 3 October 1809, nos 44–6. In her letter, Manni Begam explicitly denied Soorteellee's, and the other concubines', claims either to the stipend or to the material goods of Rustam Jang, by the Shia doctrine. She thus found a ground for claiming the whole which had to be acceeded to by both Sunni and Shia schools. In the mahrnama she submitted on behalf of Shahrukh Begum, no mention was made of Shahrukh's birth-mother, Mohumdee, herself a daughter of a concubine of Miran's, a fact that would have acceded the right of Miran's male heirs to Shahrukh's share. She submitted a fatwa under the seals of Muhammad Rubbee and Ghulam Survur upholding the payment of mahr-i-maujjul (dower payable on demand).

[83] AGG Rocke to G. Swinton, 26 March 1810, and 6 April 1810, BPC, 10 April 1810, nos 46–9. Rocke's list of 'eight females and two slaves' is a perfect example of the androcentric assumptions that went into defining slavery, discussed in Introduction.

[84] Pattle to Monckton, 1 July 1809, BPC, 8 July 1809, nos 57–8 and enclosures.

each child had been born from, the Superintendent asked both the sons and Rahīman to submit a list of the claimants. One list submitted by the adult sons of Shahamat Jang counted seven women under the category of widows by nikah, and ten women as *khadimah*. The Superintendent of Nizamat Affairs, doubtful that the former could 'produce any proofs... to sustain their assertion of marriage contracts having passed between their deceased father and mothers', compared this with the second list submitted by Rahīman, who claimed to be the legal heir, and found that all the women of the first list characterized as kaniz.[85] Three of these 'illegitimate' sons, all older than the five-year-old Bakir Ali, born of Rahīmunnisā, came to the attention of the British authorities because the Nazim Nasir-ul-mulk complained that these sons 'by concubines' had obstructed the agents of his 'sister' Rahīmunnisā, when they had tried to hold the *punyah* in her name.[86] Many of the lands acquired by Shahamat Jang were found to be registered in the names of these older and slave-born sons—Saiyid Ali, Babar Ali and Ja'far Ali; the latter took possession of their father's lands upon his death.[87] As Pattle summed it up to the Collector, the Government, called upon to 'make provisions from the Funds of the Nizamat for the support of' these women and children, possessed 'too great an Interest in the application of the property of the deceased to its just purposes to abstain from the exercise of interference on that point'.[88] On this ground, the attachment of all the landed property in dispute was ordered. The government of Bengal too wrote to the Court hoping that the stipend it had been forced to disburse to this dāmād could be henceforth resumed from a 'family which possessed so large a property... even altho' the stipend of Shehamut Jung... had been declared hereditary' in 1796.[89]

[85] The correspondence is summed up in BC F/4/312/7131, and Table I, Appendix I.

[86] Pattle to Monckton, 11 July 1809, BPC, 5 August 1809, no. 79.

[87] Pattle, Superintendent of Nizamut Afairs to Persian Sec. Swinton, 19 August 1809, BPC, 10 October 1809, no. 54. According to the amins, the taluqs, which included sub-leases and lakhiraj land, were rated at Rs 42,933–12–15–1, and paid revenue amounting to a sum of Rs 25,000. By a different computation, the mass of wealth, including houses, jewellery, clothes, and revenues from land in the holding of Shahamat Jang amounted to four or five lakhs. *See* Appendix I, Table I.

[88] Pattle to D. Campbell, 14 August 1809, ibid., no. 55.

[89] Poll. letter to Court of Directors, 18 October 1809, BC F/4/312/7131.

By 23 October 1809, (approximately two months after Mȧnni Begam had demonstrated the success with which such claims might be made), Rahīman put in her claim upon the entire wealth of her deceased husband on the basis of her marriage settlement (mahrnama) of seven lakhs. Forced to accept a principle of 'Islamic law' that claims of mahr had to be satisfied before any division of an estate, the Government of Bengal then transferred Rahīmunnisā's stipend of Rs 250 per month to the maintenance of the nine 'illegitimate' sons and daughters and sixteen concubines and female slaves, and ordered the complete resumption of Rs 1000 of her deceased husband's Nizamat stipend into the Collector's Treasury. Thus the Company's acquiescence in a decision on dower based on Islamic law allowed the Government in Calcutta to withhold the cash allowances.[90] In 1811, when another two such dāmāds died, disputes about marriage claims were avoided by documents which unequivocally left 'the control of all affairs', and the dependants, of the deceased male to Manni Begam's superintendence, specifying that the division of estates would occur according to the 'law of the Twelve Imams'.[91] In these instances too, the English Government selected different parts of Islamic law. Where Manni Begam recommended a division of the stipend which gave precedence to the slave-born and only son of a deceased man, allocated more than half of the stipend for the other slave-born children as well as allowed a deceased man's sister to inherit one part of his stipend, the Government brought in doctrines of legitimacy and illegitimacy to reduce the amount granted to the 'illegimate' children and denied the sister's claim completely on the grounds that she already received a stipend as a widow in another household.[92]

Litigants' ability to cite principles of Islamic law depended on particular sets of circumstances on the ground. Just as Manni Begam's ability to manipulate events and factions, including the Agents, had led to successful claims for dower, Rahīman's ability to claim this

[90] Pers. Sec. Monckton to Suptdt. of Nizamat Affairs, 5 July 1811, BC F/4/372/9259.

[91] Trans. Will of Nawab Sayed Bakir Ali Khan deceased, dtd. 28 April 1811, in letter from Manni Begam recd. 25 August 1811, BC F/4/395/10033.

[92] Pers. Sec. to Suptdt. of Nizamat Affairs, 27 September 1811, ibid., and AGG Cobbe to Sec. GOB, Macnaghten, 29 June 1833, Extract Poll. Cons. 11 July 1833, no. 144, BC F/4/1475/57976.

ground depended on the strength of her brother and mother within the power-groups of the Nizamat. As an official noted, her own brother, Mīr Mehdi and their mother Lutfunnisa, supported her legal battle, which consumed all the resources in litigation, as a result of which she became 'an object of dislike' to her immediate kin, and extremely dependent on her half-brother, the Nazim.[93] The latter's resources may have been crucial to Rahīman's legal success. If a particular daughter belonged to a faction at that moment out of favour either with the Nazim, or with the Agent, her ability to claim anything at all was seriously jeopardized, Islamic law notwithstanding. Nūrunnisa, another of the daughters of a Nazim, found this out for herself when one de'orhi head, Amīrunnisa Begam, took up cudgels on her behalf against the husband with whom she had 'long been on a footing of bitter enmity'.[94] However, the fact that this dāmād, Mīr Mohsin Salabat Jang, was favoured both by the Agents and by the Nazim Humayunjah, both keen to curb Amīrun's power, meant that Nūrunnisā's chances of gaining anything at all were slim from the outset. In 1837, in an effort to repay her debts and incensed by the fact that all the wealth accumulated by Salabat Jang as a dāmād was being gifted away, Nūrunnisā's Diwan filed a marriage-claim for Rs 60,000. The Agent, the husband and the Nazim combined forces to crush the suit as 'contrary to the usages of the Nizamut'. Clearly, there were two phases in the Company's policies in applying Islamic law to the Nizamat. In the first phase, which coincided with the dominance of Manni Begam and lasted till her death in 1812, officials framed issues in terms of the application of particular schools of canonical law. This framing obscured the Company's own stake in the ways in which particular schools of 'Islamic' law were selectively privileged while resolving conflicts involving substantial wealth. In the second phase an uneasy compromise between law and custom was made, which witnessed a complete marginalization of the very 'Islamic' law fashioned earlier by English officers and Agents.

In addition to the invisible matriarchs and slave-masters, and intra- and inter-generational differences, disputes about the inheritance of the slave-born and inheritance from slaves were conditioned by the

[93] AGG to Pers. Sec. Prinsep, 25 June 1821, BPC, 7 July 1821, no. 78.

[94] AGG Melville to Stirling, 14 April 1827, BC F/4/1328/52500.

acquisition and locations of female slaves. Their circulation between first acquisition and subsequent destinies (like the circulation of eunuchs between Nazims and different de'orhis) made the claims of the first acquirer and subsequent holder/ user difficult to distinguish, while their simultaneous insertion into twin reproductive cycles could yield children reared as social heirs of her master/mistress, and not as the female slave's heir. These twists and turns made the contests between their own progeny and the heirs of former masters and mistresses impossible to resolve, as we can infer from three different judicial cases from the same zamindari of Attiya in Mymensingh over 1798-1844.[95] English judges' doctrinaire absorption with legitimacy and ackowledgements of paternity obscured what were in fact intra-household contests about the limits of slave-incorporation, the heirship of the slave-born and heirship to slave-acquired wealth. For the slave-women, especially when they were mothers of children, the decisions meant a continued subordination to matriarchs like Manni Begam, or free-born heirs of masters. For the slave-born sons and daughters, it meant that instead of 'inheritance' they received 'maintenance' which remained vulnerable, in turn, to further fragmentation and resumption. The long legacy of such selective decisions can be illustrated with reference to one of the slave-born sons of Rustam

[95] Fyaz Ali Khan vs. Mussummaut Fatima Khatoon, S.D.A., I, pp. 357–8; Khajah Hidayat Oollah and Rai Jan Khanum, Moore, *Indian Appeals*, III, pp. 295–303. These disputes suggest that two different factions, led by two different wives of the zamindar Khoda Nawaz Khan, were bringing suits against each other in the name of various slave-born sons and daughters in each generation. In the first suit, Khyrunnisa who nurtured Fatima Khatoon, the daughter born of her slave Soobhani and her husband, appears to have been the major beneficiary of a district court decision in the daughter's favour. In the second suit, Beebee Ali Jan aka Pullee, the slave of Ramzan, the mother of Fyz Ali Khan, brought a claim against Fyz Ali Khan for a share of the zamindari on behalf of her son, Rujub Ali. Judges of the Sadr Diwani Adalat gave decree against the sonship of Rujub Ali. Then another suit was brought by Raijan Khanum, the first wife of Fyz Ali Khan, against his second wife Shumsunnisa Begam, for a share of the zamindari for her own son, Saadat Ali Khan, and won. The decision of the Privy Council in the last suit was in turn submitted as defence in a suit between the concubine-born daughter Noorunnisa Begam, and Shahrook Begam, the widow, of Shahzada Mahomed Kykobad of the Mysore Princes, *see* Rook Begam vs. Shahzada Walagowhur Shah, *Sutherland's Weekly Reporter*, III, 1865, pp. 187–90.

Jang, Mīr Muhammad Ali, who received Rs 50 per month as one of the aqriba. He died in 1832, leaving behind a 'lawfully wedded wife' Ashrafunnisa, two slave-concubines and a half-sister, Nurjahan Begam, the daughter of Shahrukh Begam. Though the widow claimed a share of her husband's stipend, she was denied it on the grounds that she had always lived apart from her husband and been maintained by her father.[96] Thus denied her inheritance once (at the behest of the Nazim Humayunjah, who upheld the claims of Nurjahan Begam to inherit authority over the slaves and concubines of Mīr Muhammad Ali and Rustam Jang) she was to be denied again when she claimed a share to Rustam Jang's childless, Abyssinian, slave-concubine's (called Mehbūba) stipend in 1837. When the latter died, her monthly stipend of Rs 20 was claimed by this widow and a 'stepdaughter' (Nurjahan);[97] both were refused, and the sum resumed. The Government again reiterated the principles upon which such resumptions were to be undertaken: 1) the pensions of 'mere servants' were to be resumed wholly on their death and annexed to the deposit fund and b) stipends of relations of the Nazims were to be continued to their heirs; when there were several heirs, and one or more of them died without heirs themselves, their portions should lapse to the deposit fund unless there were dependants 'the support of whom would be a duty incumbent upon the surviving heirs'. In such a case, a portion of the lapsed stipend/allowance would be made over to the surviving heirs who undertook the care of such dependants'.[98] While the Governor-General's Council belatedly allowed the provisioning of the elderly slave-servants and concubines of a collateral household, the AGG upheld the inheritance rights of the poorer heirs of the dead slave's master (the son's widow and his half-sister). He urged that 'the benefit to be derived from the resumption of small stipends enjoyed by individuals who did not have heirs of their body instead of being distributed among his natural needy heirs does not appear very decided'.[99] In the post-mortem division of some of these really small

[96] AGG Cobbe to Ch. Sec. Swinton, 16 June 1832, BC F/4/1475/57976.

[97] AGG Caulfield to Sec. Prinsep, 12 July 1838, BPC, 18 July 1838, no. 9.

[98] Sec. to Govt. Prinsep, to AGG Caulfield, 18 July 1838, BPC, 18 July 1838, no. 13.

[99] AGG Caulfield to Sec. GOB, BPC, 8 August 1838, no. 7.

stipends, principles of 'inheritance' law were pitted most visibly against the fiscal interests and socio-legal ideologies of the Company. The Government responded that Nizamat stipends were not 'heritable property' and hence could not be presumed upon by the heirs or dependants of a deceased stipendiary.

Apart from excluding such stipends from heritable 'private property', the Government of Bengal also tried to segregate immoveable from moveable wealth. These subtleties were as damaging to the slave-recipients and users of such lands and houses as they were to their poorer masters and mistresses. Regulation XIII of 1813 authorized the Magistrate of Murshidabad to raise funds for the *choukidari* police establishment and hence to levy an assessment on those slave-servants, concubines and dependents who lived outside the killah. The Superintendent of Nizamat Affairs argued in vain that the means of the greater number of these dependents being no higher than three rupees and twelve annas per month, was on par with the income of the labouring classes, 'barely placing them above indigence', and the average rate of assessment paid by these classes was equivalent to an 'income tax of three and one-third per cent... too heavy a burthen'.[100] While on one hand such slave-servants and concubines were hard-pressed by these measures, they proved to be equally hurtful to some of the less powerful and poorer heirs of masters and mistresses. While the latter had to pay similar taxes, they were barred from exercising rights of inheritance as heirs of deceased slaves. The operation of these twin-blades of colonialism—fiscal fragmentation and resumption procedures—in the case of the Nizamat thus provided the specific logic for the equalization of slave and free by legal statute in 1843. Both could be equally impoverished and immobilized. The well noted inclination of the heads of households to consume the money paid out by the Collector's office in the name of the slave thus occurred in the aftershock of this drive to equalize slave and free, and freeze processes enabling some slaves to climb into kinship structures.[101]

[100] Brooke to E. Impey, 4 May, 1814, BPC, 23 June 1814, no. 116.

[101] For the case of a maid servant and a concubine both with the same name, Pyari, in the household of a grandson of Mubarak ud daula I, *see* AGG Mackenzie to Sec. to GOB, 22 May 1860, BPC, June 1860, nos 45–8; Memorial of Syed Khoorshed Hossein to the Lt Governor of Bengal, 19 December 1860, BPC, January 1861, nos 80–1.

Noticeably absent in such decisions by mid-century was the particular school of Islamic law—Shia or Sunni—the English judges and officers were proceeding from. This affected many female slaves, their progeny, and their masters and mistresses' heirs because wala was differently accomodated in different schools of law. Particularly relevant to the claims of women who had begun as slaves like Babbu, Manni and Faizunnisa (Walida) Begam in the household of the Nazim, or women like Mehbuba and Harun in collateral households, the question whether having borne children to their masters, such females also acquired the capacity to inherit from them, or bequeath the material wealth they had enjoyed and accumulated to their slaves and children, was not judicially determined till later in the nineteenth century. In the first half of the century, these decisions were taken and implemented by officers of the Company with very little attention to the details and degrees of kinship within which emancipation by child-bearing and inheritance rights of the slave-born were articulated in different schools of Islamic law. Evidence exists of a two-fold division by which inheritance rights were to be adjudicated and implemented. In one division, were the claims of Nazims upon the wealth of the gaddinashin begams, and in another, the claims of slave-concubines and slave-born children among the inferior and collateral kin. While the latter group was rapidly shorn of the vestiges of right, the former prospered.

In the former group, the precedents established after the decease of Babbu Begam were studiously followed, regardless of genealogical kinship and Islamic law. Hence, when Walida Begam died, the whole of her estate was described as her 'personal property' and handed over to the reigning Nazim, despite the claims of one of her daughters Moti Begam, and her children.[102] Certain genealogical principles were either disregarded or rendered ineffective in such contests. In Babbu Begam's case, the claimants in her 'property' did not include the blood-relatives of Saman Ali Khan, the putative father of Babbu Begam, heirs who were alive at the time of the Begam's death. In the case of other widows of previous Nazims, like Bahu Begam who died in 1849,

[102] AGG Ricketts to Actng. Pers. Sec., 29 November 1822, BPC, 7 February 1823, nos 54–6, corroborated in later evidence, in Subordinate Judge of Murshidabad BC Sil, 7 June 1872, NAI, FPP, A, October 1877, no. 525, Exhibit F.

they neither left genealogical children of their own, nor were their natal kin, having been servants in the de'orhis, in a position to challenge the Nazim's heirship. In 1858, however, when Amīrunnisa, the gaddinashin begam of the Company's making, died, the consequences of Amīrun's own biological childlessness, compounded by the hollowness of claims based upon being 'nurtured' by her, came to the fore. The Nazim's claim to inherit all her property was challenged by her half-brother, Saiyid Mehdi Ali Khan. She had reared this son of her father and his concubine as 'her own son' from his infancy,[103] perhaps as an attempt to replicate Manni Begam's rearing of Qaim Jang, and in very similar conditions of conflict with the regnant Nazims. This brother claimed under the operation of the ordinary Islamic law of inheritance, succession to his elder half-sister; the Nazim claimed 'not as the natural heir of the Begam under the law of inheritance, but by virtue of a family custom which gave it to the person holding the dignity of the Nawab Nazim'.[104] Where previous Nazims had inherited from women who had begun as slaves and then become mothers and grandmothers, Amīrun was certainly *not* a first-generation bāndi. Nevertheless, faced with the history of the Company's involvement in adjudicating these conflicts in the earlier cases, the Government of Bengal allowed the Nawab Nazim to take 'possession' of the properties in contest, while submitting Mehdi Ali Khan's claims (and subsequently that of two of his daughters) to the courts.[105] The courts validated the 'custom' of the Nazims over and against the claims of 'natural heirs' of the brother of Amīrunnisa. When the final gaddinashin, Ra'isunnisa died in 1893 the issue was reopened again. Unlike Amirun, Ra'isunnisa had been a first generation slave-girl belonging to Moti, and had been taken into the haram of the Nazim after the birth of her son. When she died, she left behind many who were, in biological terms at least, her 'grandchildren', i.e. the many sons and daughters of Faridunjah alone, apart from the heirs of her daughter Gītiāra Begam. Thus the Nawab Bahadur's claim to the sole heirship to the two lakh-worth of cash, jewels and moveable

[103] Petition of Saiyid Mehdi Ali Khan, WBSA, GPP, 16 June 1859, no. 2.

[104] Commissioners under the Nawab Nazim's Debts Act, 10 May 1875, BPP, June 1875, no. 11.

[105] The decrees of three courts in Omraoh Begum and Zohra Begum vs. Nawab Nazim of Bengal are found in NAI, FPP, October 1877, nos 519–31.

properties, and the zamindaris yielding an annual profit of Rs 12,000 was certainly impossible on the basis of a 'natural' heirlessness, as had been presumed in the case of Amīrunnisa. The other grandchildren, half-siblings of the Nawab Bahadur, well aware of the 'dependent and comparatively helpless position of other members of the family' *vis-a-vis* their more powerful relative, based their claims on the ordinary Shia law of inheritance.[106] Since under the indenture of 1880 and the subsidiary Act XV of 1891, the Nazim and his eldest son had specifically given up all rights, privileges and claims attached to the office of the Nazim, the Nawab Bahadur based his claim solely upon 'family custom' and upon his rights as a 'head of the family'. Earlier political privileges of 'Asiatic princes' gave way before the domestication of authority. In the garb of the private patriarch, the Nawab Bahadur defeated the claims of his junior kin. In the process, two legal distortions occurred. The differences between slave and non-slave woman were erased and the emancipatory potential of becoming a mother was ignored at the same time that claims between siblings based upon the Shia law of inheritance were also defeated.

Heirship from Eunuchs

The Company's government, and its courts, were clearly complicit in the conflicts between heirs of slaves and heirs of masters over wealth which slaves administered and from which in turn they were maintained by the household, especially in the case of important and visibly powerful slaves like the eunuchs. The issue of the heirship to these slave-officials, whose death had serious implications for the household and treasury, was flushed out particularly with the imminent death of Basant Ali Khan, an aged eunuch, in 1833. Finding the house he lodged in surrounded by guards sent by the Nazim, led by the Nazim's own eunuch, Firoz, with orders to put seals on his house and property 'so as to prevent his even having access to his Linen, food or medicines', Basant appealed to AGG Cobbe. The latter then wrote to the Superintendent of Nizamat Buildings resident in the killah, Colonel McLeod, to personally remove the seals and guards, and to ensure the drawing up of a will by the infirm eunuch. Whether the

[106] Memorial of Vahid Ali Mirza and others, WBSA, Judcl., November 1893, no. 88.

papers were actually drawn up and transferred to the Agent we do not know, but McLeod reported that the invalid 'stated his intention to bequeath all his landed possessions to the Company, with a request that a provision therefrom of comparatively small extent, should be made for the support of his dependants... that he meant to make over the Kuddum Shurreef to the Bahoo Begam &... his personal property to his friend (Bhaii) Bahr Ally Khan'.[107] This disposition surprised the Englishman because the eunuch 'did not once mention' the mistress he had served so long, the Dulhin Begam, Amīrunnisa. The 'brother' the term Bhaii refers to was also another eunuch, belonging to the de'orhi of Manni Begam. On what ideological or economic bases did such civil servants attempt to adjudicate the inheritance of the 'possessions' of the eunuchs after their decease? The Agents and the Company's decisions fluctuated according to the amount of the wealth purported to have 'belonged' to the slaves. Thinking that the eunuchs *possessed* the wealth they managed, successive AGGs tried to devise testamentary evidence from Basant Ali Khan himself as to the modes in which the wealth would be bequeathed upon his death. Perhaps, the eunuch recognized the possibilities that the Agents represented of dispensing with something like his 'self-acquired' wealth, and also realized that the cupidity of the Company's officials would ensure the success of even the tiniest act of defiance. The AGG, in 1833, at least believed that Basant, by his will, would leave to the Company 'about 3 lacs of Rupees the provision for his followers will involve a very trifling expense'.[108] However, in the paper that he wrote at the behest of the AGG, Basant Ali left the 'whole of his property moveable and immoveable' in trust to Bahu Begam and Bahar Ali Khan who were directed to appropriate two-thirds to the support of the Qadam Sharif Masjid and the Imambara at Begam Ganj, and the remaining third was to be used for his own funerary expenses. The supervision of the Government in the expenditure of the funds was requested, especially if 'opposition be offered in any quarter'. The AGG apparently remonstrated on the 'difference' of the above disposition 'from that

[107] Coll. D. McLeod to AGG Cobbe, 10 June 1833, BPC, 21 June 1833, no. 70, BC F/4/1475/57976.

[108] AGG Mcleod to Sec. to GOB, Macnaghten, 12 June, 1833, BPC, 21 June 1833, no.70, ibid.

which was understood some time before was to be made'.[109] Even though the ailing eunuch clarified that there was no discrepancy, the sums mentioned in this clarification were substantially less than the three lakhs the AGG had looked forward to. In fact, the total annual income from the zamindaris were calculated by the eunuch to be between Rs 8000 to 9000. An aggrieved AGG Thoresby remarked that Basant Ali Khan had written nothing of the 'small houses and some Gardens and patches of ground' in Murshidabad, a statement of which he tried to procure from the *mutsaddis* of the deceased eunuch. As far as the English were concerned, the 'proprietorship' of these zamindaris, was adequate criterion for them to insist that Basant Ali Khan was 'a free servant of the Nizamut'[110] (in direct opposition to the Nazim's claims). This arbitrary ascription of 'freedom', in turn, was meant to establish the ability of the eunuch to dispose of his wealth, in ways that evaded the erstwhile masters.[111] Prinsep, an official of the Government of Bengal, who vehemently opposed any legislative enactment on the abolition of slavery, appeared to agree with the majority of the Law Commission appointed in 1837 on disallowing masters (like the Nazims) from inheriting from their slaves. The specific instances Prinsep used to illustrate his argument were those of the main eunuchs of the Awadh and Nizamat households.[112]

Since the Company's officers treated all acquisitions and divestments of individual members of the Nizamat as though these were in 'State' property, the dilemma for the Company's officers lay in balancing Company interest in revenue alongside adjudicating a 'private' inheritance contest in which there were too many claimants.

After the death of Manni Begam, Basant Ali Khan though receiving his allowance of Rs 350 held no office, until Amīrunnisa's elevation to the headship of the De'orhis. Since Amīrunnisa then re-appointed

[109] AGG Thoresby to Sec. to GOB, Macnaghten, 13 January 1834, ibid.

[110] Ibid.

[111] For identical attempts to resume lands 'gifted' by Walida Begam to her eunuch Roz Afzūn, *see* letters of Cobbe and McLeod to Nawab Nazim, 16 June, and 7 September 1835 and 9 March 1836, *MNLI*, II, pp. 9, 14, 23.

[112] Minute of 31 July 1841, BC F/4/1947/84542. Prinsep discussed the differences in British policies: on one hand, allowing the Nawab of Awadh to claim as heir of the eunuch-slave of his grandfather, Dārāb Ali Khan, and on the other, refusing to resolve the claims in Murshidabad.

Basant as Nazir, a post for which he received a 'salary' of Rs 200, Amīrunnisa's claims rested both upon her succession to headship and to the perquisites of that status, one of which was the services of all those slaves who had served the earlier head. This was significant because even though at the end of his life, Basant Ali Khan suffered from palsy and was unable to actually conduct the affairs of his office, and another eunuch, Zamurrud, was the real functionary, the office and title of Nazir, along with the salary, remained with the elderly eunuch. Yet Basant had not mentioned Amīrunnisa Begam who had been his employer for the last decade of his life. This was not as revealing as the Agents thought. Nor had he tried to bequeath anything to his own chelas, Miyān Jawahar and Miyān Kambaksh, nor had he mentioned Zamurrud, the slave he had personally bought from Hyderabad. More than their biological heirlessness, Basant Ali Khan's naming of Bahār Ali Khan and Bahu Begam as trustees of a *waqf* that he set up clarified that even if they left behind their own acquired 'chelas' with the potential to act as 'legal heirs', they did not try to pass on heirship to their own slaves. Perhaps the eunuch had not mentioned either because it was taken for granted by him that title to various houses, lands, gardens were not his to dispose of. In fact, in the very people he named as trustees lay the indication of this knowledge: Bahār Ali Khan was the eldest eunuch of the Nizamat, and served at the de'orhi of the Bahu Begam. Though appointed a trustee of the estate of Basant, he never actually exercised this power, possibly because of factional machinations which caused him to resign: by 1838, Bahār too had died.[113] Initially, the Government of Bengal upheld the testamentary evidence, on the grounds that Basant Ali Khan 'appeared to have been a Free person having a right to dispose of his own property'.[114] Five years later, it changed its mind. After taking over the charge of the estate and appointing another manager had failed to yield the expected profits and revenues,[115] Government

[113] Melville to Macnaghten, 1 May 1837, *MNLI,* II, pp. 84–5, 157.

[114] Extract Poll. Letter to Court of Directors, 13 November 1834, para 230, BC F/4/1522/60090A. One of the factors influencing the decision was the fact that 'several of the villages in default of payment of the Government dues were advertised for public sale'. See para 228, ibid.

[115] For 'Abstract Accounts of Receipts and Disbursements from 1240 B.S. to 1243 B.S.[1834–5 to 1837–8]' furnished by Bahār Ali Khan, *see* BPC, 13 February 1839, no. 7.

accepted that 'Bussunt[sic] Ali Khan was a purchased Eunuch slave his heir therefore, failing any near relations or dispositions by will of the Estate, will be, according to the Mahomedan law, his late master'.[116] Finally, the issue of handing back the estate to the Nazim or to the Begams was reopened. But after 1839, there was a new and very young Nazim on the masnad (ten-year-old Faridunjah), who was deemed 'unqualified for such a charge'. So, despite the fact that the eunuch's deposition had specified Bahu Begam, the charge of the estate was handed over to Amīrunnisa Begam. The conformity of the Company to formal principles from sharia or to a slave's testamentary provisions were both conditioned by the Company's fiscal and political interests.

The choice of Amīrunnisa Begam was not an accidental one. Capitalizing on the proclivity of the current Agent, to believe unproblematically in the 'dissoluteness' of all females in the Nizamat, and continuing a tradition of politico-symbolic gestures, the Nazim Humayunjah had accused his grand-aunt Bahu Begam of 'sexual misconduct with a Chela of the Nizamat named Shooja Koolee alias Hingoo'.[117] After a summary investigation, involving depositions by the eunuch Bahār Ali Khan himself and other slaves of the Nazim, Melville authorized the removal of the 'grey headed old and nearly toothless' Bahu Begam from the de'orhi from which she claimed her rank. Though a subsequent Agent showed how 'impossible' this charge was, it was clear that one of the reasons for the Nazim's conflict with this elder female kinswoman (uncle's wife) was another concurrent battle over inheritance.[118]

As we have seen in the instance of the slave-born sons of the dāmāds, or the brothers and sisters in conflict with various Nazims, or that of eunuch trustees of eunuchs, the desire to harness the wealth of masters involved the Company and later colonial administrations in the same acts of denial of the 'natural heirs' of slaves as the slaves' masters demanded. This culminated in multiple and growing paradoxes. For one, the denial of masters' claims in the wealth slave-

[116] H.T. Prinsep, Sec. to GOB, to AGG, R.B. Pemberton, 13 February 1839, *MNLR*, II, pp. 384–5.

[117] For moves against women charged with adultery, *see* Chapter 2, and Nawab Nazim to Melville, 24 June 1837, BPC, 25 July 1837, no. 7.

[118] AGG Caulfield to Prinsep, 26 April 1838, and enclosures, *MNLI*, II, pp. 171–4.

eunuchs donated to the cause of a higher being came about in the first half of the nineteenth century only by a suppression of the Islamic law (of wala) the Company had vowed to uphold. Faced with falling revenues from the same properties earlier managed by the eunuch, the claims of previous masters were resurrected and accepted. What was considered Basant Ali Khan's estate was made up of *taluqas* and zamindaris in Purnea, Murshidabad, Bhagalpur, the total annual income from which amounted to Rs 19,760 by 1861: from this, government received as revenue Rs 9370, and a major portion of the remainder was expended on the upkeep of the Qadam Sharīf.[119] Beside this mosque, with its stone impression of the Prophet's foot sacred to the Shi'a populace of the city, lay interred both the grand eunuchs Basant and I'tibār.[120] Establishing control over the revenues of such mosques, evident in the case of other mosques like the one founded by Manni Begam[121], drove some of the resumption procedures and may have equally influenced Act V of 1843. In effect, Government's assent regarding Amīrunnisa's management of Basant's estate continued the superintendence of Dārāb Ali Khan, the eunuch of the Nazims Humayunjah and Faridunjah, till 1867, when under the terms of the Act XX of 1863, a three-member waqf board was constituted.[122]

The paradox deepened in the second half of the century, when put to the test in colonial courts of law, both colonial legal opinion and administrative action around slave-held wealth were compromised. As in the instance of the property of the gaddinashin Begams which was allotted to the Nazims as 'family custom', the appropriation of eunuch-held lands and wealth by the Government as 'State' property led to eunuch legatee, Miyān Arjumand appealing to the Government of Bengal for the payment of the allowances which Nazim Faridunjah had promised to the estate of the eunuch Amān Ali Khan from the

[119] AGG Thomson to Sec. GOB, 30 May 1862, WBSA, GPP, July 1862, no. 29.

[120] For a description of the architectural style of this complex, begun in 1780 by I'tibār Ali Khan, Manni Begam's chief eunuch, *see* Perween Hasan, 'Art and Architecture' in Sirajul Islam, *History of Bangladesh*, III, p. 664.

[121] For advice to establish control over the revenues while omitting the care of the slave-servants fixed to the mosque in Chouk Nowabad, *see* Macnaghten to AGG, 21 June 1836, BPC, 19 June 1836, no. 3.

[122] Collr. Murshidabad to Commr. Presidency Divn., 3 June 1896, WBSA, Judcl., September 1896, no. 29.

revenues of Rokunpur.[123] Since the parganah of Rokunpur also counted as 'State' property under the terms of the indenture signed by the Nazim, the allowances payable to the waqf of Amān Ali Khan (Golkothi) in Calcutta had also been stopped. The question really boiled down to who should pay the allowance, the Government of India or the Nawab Bahadur Ali Kadr, who was considered to be the 'head of the family'. After some wrangling between legal opinion about the effect of the abdication of the Nawab Nazim on contracts and deeds made antecedent to his abdication, the Government of Bengal decided to pay the sums 'from the Indian revenues as a pensionary charge'.[124]

As for Basant Ali Khan's waqf, though Government derived revenues from its management, by 1878 when an enquiry into the availability of funds from 'Muslim' endowments was instituted, it was found that the 'estate was encumbered with a heavy debt, and had to conduct expensive lawsuits brought against it'.[125] How had matters come to such a pass in an institution cherished by slaves, and non-slaves alike, and managed under the authority of the colonial government? The President of the Committee appointed by the Government in 1867, Saiyid Azam Ali Khan, died in 1896. The Nawab Bahadur of Murshidabad (heir and successor of Faridūnjah) claimed the sole management of the estate and mosque on the grounds that he had inherited the heirship to the slave's estate from both his father and grandfather and from the gaddinashin Begam entrusted with its management by the government. Half a century after the passing of Act V of 1843 the Nawab Bahadur claimed that a) the Act had not destroyed his right to slave-property established prior to the Act, and b) the management of the estate by Dārāb Ali Khan, another eunuch, did not jeopardize the Nawab Bahadur's claims, but instead confirmed the 'custom' by which

> the head of the family or of the de'orhi to which the eunuch slave might be attached to grant to such slave gardens and other immoveable property, and

[123] Offg. Sec. to GOB to Solicitor to GOI, 30 September 1882, BPP, January 1883, no. 20.

[124] Offg. Sec. GOB to Sec. GOI, Foreign, 25 January 1883, ibid., no. 23.

[125] Offg. Sec. to BOR to Sec. to GOB, 21 Nov. 1879, BPC, March 1880, file 19, no. 6.

to permit the slave at his death to establish for religious purposes certain endowment.[126]

The Nawab Bahadur's claim to heirship to this slave's estate was referred by the Government of Bengal for legal opinion. According to the Advocate-General of India, the Nawab was *legally* right and 'entitled to the Mutwalliship as the successor to Amīrunnisa Begam'.[127] Though Government preferred a suit to be brought in the civil courts to uphold this right, the Advocate-General had been fairly consistent in his interpretation. For in 1870 and again in 1893, the Courts had upheld the Nazim and his successor the Nawab Bahadur in their claims to succeed to the properties and privileges of the gaddinashin Begams, to the detriment of the 'natural heirs'.

Colonial policy regarding the civil consequences of enslavement and slave use was not conditioned by attention to doctrinal law, nor was it consistent. The political and economic interests of the Company gave it a material stake in making everything taxable and resumable at will. However, in constructing law, the officers and judges of the Company ignored the fact that slavery and kinship were not distantly removed poles, but a living continuum as far as some of the larger slave-holding households went. The slave of one generation could be, and often was, the female ancestor of the next: and slave-born children in turn were siblings in the same generation, or nieces and nephews, or grandsons and grand-daughters in two or more generations. Adjudicating inheritance issues in contexts of such inter-generational mobility of slave-born, compounding the inheritance from slaves by the inheritance of slaves (as property) and inheritance by slaves, would have required very different notions of slavery-as-a-state. It would have required minute examinations of the flows of 'social capital' and the languages and claims of kinship that English colonial jurisprudence had no room for. In the first half of the century, therefore, officials of the Company ignored the claims of the 'natural heirs' of deceased slaves, as much as they ignored the claims of collateral and junior kin upon the Nizamat. When the natural heirs of some slaves also joined the ranks of junior kin, their elimination from the inheritance of their

[126] Nawab Bahadur Murshidabad to Collr. Murshidabad, 30 April 1896, WBSA, Judcl., September 1896, enclosure no. 29.

[127] G.C. Paul, in Sec. Legal Affairs to Ch. Sec. GOB, 18 August 1896, ibid., enclosure no. 33.

slave-mothers was continued under different guises—sometimes as illegitimacy, and sometimes as the political right of the Nazims. Thus in the first half of the century, the differentially stratified members of the Nizamat were caught between twin blades. On one hand, the Company's policy of 'lapsed stipends' and 'resumption' proceedings in land necessitated the oversight and suppression of wala, to the advantage of the various municipal and political measures that the Deposit Fund financed.[128] On the other hand, there were escalating conflicts over the custody of particular slaves and efforts to control the funds of, and inheritance from, slaves. By the second half of the century, these myriad claims of masters and mistresses upon slaves and ex-slaves had been ideologically displaced: the substance of wala and related terms and usages were relegated from the realm of the patronate to that of family and inheritance law. Thus a peculiarity of Bengal case-law in the second half of the nineteenth century was the simultaneous proliferation of 'family' and 'inheritance' disputes revolving around slaves in the household, and an absence of pleading on any of the clauses of Act V, particularly that relating to disabilities of inheritance following from slave-status, in sharp divergence from case-law in Bombay.[129] Despite formal adherence to Islamic law, the practice of English courts and officers in Bengal altogether ignored or distorted fine distinctions between Sunni and Shia schools of law, as we have seen above. These contortions were in turn bequeathed to later English judges and officials as precedent, the net result of which was a parting of ways between case-law and doctrinal Islamic law;[130] as also the consolidation of something like 'customary' practices within

[128] Apart from the palace, upon which the sums spent appear incalculable even by the Government's own admission, for an example of 'public works' financed from the Deposit Fund *see* IF&P, 7 January 1859, nos 6–7.

[129] Wala was explicitly pleaded in Sayad Mir Ujmuddin Khan, legal representative of Fatma-ul-nissa Begam vs. Zia-ul-nissa Begam and Rahimulnissa Begum, ILR, 3Bom, 422. The Privy Council decision of 17 March 1879 dismissed the claims of heirs of emancipators of slaves and upheld the inheritance rights of the natural heirs of slaves.

[130] *See* Ranee Roshun Jahan vs. Rajah Syud Enaet Hossein, nos 158 and 178 of 1865, *Sutherland's Weekly Reporter*, V, 1866, Civil, p. 17. In appeal, the Privy Council upheld the High Court in 1876, *see* Khajoorunnisa, widow of Enayut Hossein vs. Rowshan Jahan, ILR, 2Cal, 1877, p. 184.

slave-holding households. The manipulation of specific doctrines from different schools by English judges and officials was even more explicit in what had become by mid-century the Company's legal regulation of slavery.

5

Legal Complicities: Transactions in Slaves and Company Regulation

The intersection of slave-laws of two different historical formations in the late eighteenth and nineteenth centuries is sadly understudied in the present scholarship of the reconstitution of 'Islamic' law in India, and of indigenous law in general.[1] For the purposes of the Company in Bengal, there were three inter-related but specific issues at stake. One was the right of masters in their slaves extending to the aid of the state in effecting the recapture of fugitives. The second was the ability to garner powers of terror and coercion adequate to 'disciplining' slaves. The third was the protection the law offered to slaves. The officials of the Company, though supposedly working alongside indigenous jurists, actually regarded

[1] *See* Upendra Baxi, 'People's Law in India: The Hindu Society' in Masaji Chiba (ed.), *Asian Indigenous Law: In Interaction with Received Law* (London, 1986), pp. 216–66; Michael R. Anderson, 'Islamic Law and the Colonial Encounter in British India' in Chibli Mallat and Jane Connors (eds), *Islamic Family Law* (London, 1990), pp. 205–23; and 'Work Construed : Ideological Origins of Labour Law in British India to 1918' in P. Robb (ed.), *Dalit Movements and the Meanings of Labour in India* (Delhi, 1993), pp. 87–120; J. Fisch, *Cheap Lives and Dear Limbs: the British Transformation of the Bengal Criminal Law 1769–1817* (Weisbaden, 1983). Despite their significant contributions, these scholars say very little about the adjudicatory norms regarding slaves and slavery in the dialectic of received and indigenous law. The exception is Radhika Singha, 'Making the Domestic More Domestic: Criminal Law and the "Head of the Household"', 1772–1843, *IESHR*, 33, 3, 1996, pp. 309–43 and *A Despotism of Law: Crime and Justice in Early Colonial India* (Delhi, 1998) pp. 152–67.

the first of these as the primary object of all regulation. From the way the decrees and statutes were worded, the issue of protection to a slave became synonymous with protection of a master's claims in a slave. To protect the rights of the master over his/her slave in turn became part of the protection of the domestic market against the encroachments of other foreign powers. In the process, the site of the 'domestic' was inscribed into law twice over, once as the antonym of foreign (in terms of territorial jurisdiction and genealogical origin), and then, as the authority of the master of the 'family'. Instead of a separation between police regulation of slaves and regulation of commerce in slaves, the entire body of Company regulation operated simultaneously to reinforce the validity of both kinds of 'domestic' slavery—one within the household, and the other within the Company's territories.

Protecting Domestic Slavery: Company Regulation and Foreign Traders

The fact that most of the Company Regulations of the late eighteenth century were directed against the French, Dutch and Portuguese powers in India has escaped the notice of historians and led to a confused assessment of Company regulations with regard to the slave trade. For instance, Cassels has argued that because 'Muslim law had superseded Hindu law in the Company's criminal courts', it was possible for the Government of Warren Hastings to demonstrate respect for Muslim law and yet at the same time to take a stand against the practice of stealing children from their parents and selling them for slaves.[2] Based upon a misreading of historical semantics and of 'Islamic law' in the eighteenth century, Cassels' portrayal of these regulations as early attempts at abolition, 'poorly enforced', is hardly corroborated by the evidence.

From 1740, the English, both on the Continent and in the colonies were at war with the French. Even during the intermittent periods of peace between the two powers, complaints against the French monopoly of trade were aired. Particularly for English officers

[2] Nancy Gardner Cassels, 'Social Legislation under the Company Raj: the Abolition of Slavery Act V 1843', *South Asia*, 11, 1, 1988, pp. 59–88.

commanding regiments of the Company's armies in India, the French ability to muster strong navies and armies on various islands in the Bay of Bengal offered a sharp contrast to the East India Company's difficulties in recruitment, large-scale desertions by, and deaths of, the native soldiery. The recruitment of slaves for a variety of military functions had become 'universal practice' among the Portuguese, Dutch and French powers in the West Indies by the mid-seventeenth century. It was equally firmly established in the form of the British West India Regiments by the late eighteenth century.[3] With the outbreak of the American War of Independence from 1775, in which the French joined the colonists, the wars with Mysore between 1789–99, and the renewed hostilities of the French Revolutionary Wars from 1792, every success of the French and allied armies and navies in adding to their labouring groups, was feared by the English Company in India. Thomas Deane Pearse, who had served in the King's Infantry in the West Indies before joining the Company's army in Bengal attested to this hostility to French deployment of slave-lascars in 1772.[4] A letter to the Court of Directors, communicated to the Supreme Council at Fort William of 1780, suggests the Company's acute awareness of the slave-based strength of the French and Dutch forces.[5]

Hence the distinctive feature of the earliest regulation-law was the success with which the English Company managed to safeguard its own, and allied, slave-holdings from the depredations of the Dutch, the Danes and most particularly the French. In February 1774, Khan Jahan Khan, the faujdar of Hughli, reported that 'a black Portuguese named Antony' had been convicted of carrying away some women

[3] Roger Norman Buckley, *Slaves in Red Coats: the British West India Regiments, 1795–1815* (New Haven, 1979) and David Patrick Geggus, *Slavery, War and Revolution: the British Occupation of St. Domingue, 1793–98* (Oxford, 1982), pp. 315–55.

[4] 'Memoir of Colonel Thomas Deane Pearse of Bengal Artillery', *BPP*, 2, 1908, p. 316.

[5] Letter to the Court of Directors from Capt. John Buncle, Commander of the *Warren*, Cartel Ship, from the Cape of Good Hope, 20 December 1780, regarding the increase of the military force of the French, the joining of Dutch and Danish ships with the French fleet, which included 'soldiers and slaves'. *See* British Library, Hastings Papers, Add. Mss. 29199, folios 509–12.

by force. The women had been set at liberty, but the wife of Antony had instituted a suit against the imprisonment of her husband. In response, a set of regulations were passed whose aim was to reorganize the police of the town of Calcutta.[6] These removed the power of recognizing complaints of 'Christian slave-owners' from the *faujdar* to a Superintendent of Police, who was authorized to punish all slaves and servants that had deserted (Regulations 3–8). However, Regulation 9 also addressed the European trader, stipulating that he follow the

> 'ancient law of the country (which requires that no slave shall be sold without a Cawbowla or Deed attested by the Cauzee signifying the Place of the Child's abode & if in the first purchase, its parents names, names of the seller & Purchaser & minute description of the persons of both).

Following this logic, Regulation 10 stipulated that 'no person shall be allowed to buy or sell a slave, who is not such already by former legal purchase'.

From 1771, the Council of Revenue had been aware of the Portuguese proprietors of slaves who applied for the authentication of the bills of sale of slaves to the Collector's office in Calcutta.[7] In the Minute accompanying the Regulations of 1774, the targets were picked out. These 'judicious precautions' were to prevent the 'numbers of children... conveyed out of the country on the Dutch, and especially the French vessels'. While the Minute did also speak of 'abolishing the right of slavery altogether', there were hedges placed around the Englishman's rights of property ('where slaves have become a just property by purchase antecedent to the proposed prohibition') and that of indigenous masters ('the most creditable of the Mussulmen and Hindoo inhabitants'). Therefore, when the Council at Dacca enquired whether indigenous masters could keep the children born of their slaves in bondage, the Calcutta Council decided that the right of masters to the children of slaves, already their property, could not legally be taken from them in the first generation.

[6] BOR, 17 May 1774, no. 214.

[7] For an instance of a slave-holder, Antony Castano, submitting two bills of sale transferring two slave-boys to one Manoula, and for Mr. Graham's decision to make the naib Qazi who had attested the bills indemnify Anthony Castano for one slave, while allowing the latter to keep the other. *See* Narendranath Ganguly, 'A Peep into the Social Life of Bengal in the 18th Century', *BPP*, 69, 1950, pp. 46–50.

By April 1789, an editorial in the *Calcutta Gazette* referred to the extensive traffic at Kidderpore in the sale of child-slaves in these terms: '... many are annually imported from these Provinces to work for the benefit of our political rivals in other regions'. The editor then welcomed the 'effectual measures [that] will be speedily taken for the remedy of so disastrous an evil'.[8] The measure referred to was the Proclamation of July 1789 which forbade 'Europeans' from transferring slaves away from India, arranged for the pilots of such boats to be prosecuted, and offered rewards for information on such transactions. Cornwallis' letter to the Court of Directors in August, 1789, clarified that the traffic was 'carried on in this country by the low Portuguese, and even by several foreign European seafaring people and traders, in purchasing and collecting native children in a clandestine manner, and exporting them to the French islands'.[9] The Collector of Chittagong recognized that the Proclamation had aimed only at the 'transporting... and vending them [slaves] in a *foreign country*' but asked whether slaves in Chittagong should be released or whether the practice of transferring them only in the future, suppressed. The clarification of the Council left nothing in doubt. The Proclamation, it said, referred specially to the *exportation* of slaves by sea.[10]

The early Company efforts were directed towards establishing a quasi-monopoly for itself, both by prosecuting private trade and by the wresting of the trade from rival European powers, rather than the

[8] *Selections from the Calcutta Gazettes for 1789–97*, II, p. 213.

[9] Cited in G. Forrest (ed.), *Selections from the State Papers of the Governors-General of India: Lord Cornwallis* (Oxford, 1926), II, p. 140. The use of 'children' was double-edged; the records of the actual catches show that these were very young slaves, not supposedly 'free' children. For an account of Portuguese friars of Damao selling slaves to the French on the western coast, *see* Antonio Fransisco Moniz (compiled and ed.), *Noticias e Documentos para a Historia de Damao: Antiga Provincia do Norte* (Bastora, 1900–1917), IV, pp. 107–8.

[10] Collr. of Chittagong, 10 August 1789, BPubC, 21 August 1789, no. 20, and orders thereon. Emphasis added. Again in 1793, the Superintendent of Police in Calcutta referred the question of liberating, rather than restoring, slaves to the Council and was told that the 'Regulations in force relative to slavery extend only to a Prohibition against exporting Persons as Slaves, and the consequences arising from that act.' *See* Superintendent of Police to John Shore, Governor-General in Council, 18 November 1793, BPubC, 22 November 1793, no. 3.

abolition of slave-transfers or slavery.[11] Nowhere in Regulations 9 or 10 of 1774 had there been any notion that purchase *per se* was illegal. Nor, after 1789, was there a lack of enforcement: the records, in fact, show the active enforcement of regulations against the French and Danish traders between Chandernagore, and islands of the Bay of Bengal, whether far afield like Mauritius or not so distant, like Ceylon, which were controlled by other European powers.[12] A cursory glance at the names and destinations of those prosecuted confirms this.[13]

Though the English East India Company deployed a rhetoric of humanitarianism in proceeding against its rivals' slave-holdings or transfers, the underlying concerns were mercantilist. From the 1750's the Company had been supplying slaves to and from India. The obstacle the Directors of the Company put in the way of slaves being 'conveyed' away from India by individual Europeans was that of a security bond paid by the transporting holder/ merchant which was to ensure that the slave would not become chargeable upon the coffers of the Company subsequently.[14] The Directors resented the fact that owners of slaves taken to England on foreign ships did not indemnify the Company. In 1782, referring to the slave-lascars of the Danish ships sailing from Bengal and brought to England, the Directors urged that since the 'expence... for their maintenance and Cloathing... ought

[11] For similar analysis of British efforts to control the 'Arab Slave-trade' in the nineteenth century, *see* Anirudha Gupta, 'Suppression of Slave-Trade and British Imperialist Strategies in the Indian Ocean 1815–1870' in U. Bissoondoyal and S. B. C. Servansingh (ed.), *Slavery in South West Indian Ocean* (Mauritius, 1989), pp. 96–105. The broader issue—that of the relationship between antislavery and capitalism—is the subject of an ongoing debate: for the main arguments *see* Seymour Drescher, *Capitalism and Antislavery* (London, 1986); D.Eltis and J. Walvin (eds), *The Abolition of the Atlantic Slave Trade* (Madison, 1981); Robin Blackburn, *The Overthrow of Colonial Slavery, 1776–1848* (London, 1988).

[12] Thus most of the child-slaves brought from Dacca to Calcutta in the years 1787–8 appear to have been found in the houses of Frenchmen, *see* BRC, 30 January 1788, p. 511; for 53 such slaves brought by Portuguese middlemen, *see* BRC, 7 May 1788, pp. 93–4. For Indian slaves supplied to the French-controlled Mauritius, *see* Marina Carter, 'Indian Slaves in Mauritius (1729–1834)', *IHR*, 15, 1987–8, pp. 233–47.

[13] Slavery in India: Correspondence of Court of Directors and the Government in India, (*PP*), 1828, pp. 13–24, 28–41.

[14] For permission to transport slave boys and girls from India to England upon the execution of such security bonds, *see* BPubC, 12 December 1770, no. 2; 2 November 1771, no. 9; 7 January 1772, no. 11; and 25 January 1773, nos 7–8.

not to be borne by the Company' the only alternative left was to discourage the 'sending of Black Servants to Europe'.[15]

Political and commercial concerns dictated that particular foreign powers would bear the brunt of the Company's rhetoric. The outbreak of the revolutionary wars and the desire to prevent the reduction of slave-holdings influenced matters in the Company's territories in India. When in 1794, the Dutch joined the French revolutionary armies, the Council in India wrote of its fears that the Dutch would succeed in purchasing Cochin from the Raja of Travancore. This would disadvantage the Company not only because it would provide a 'depot for warlike stores' but also because it would facilitate the continuation of the 'traffic in slaves carried on to a considerable extent by the Dutch to the great detriment of the British possessions on that coast'.[16]

These assaults upon the slave-transfers of other European powers rested ultimately on the Company's need to conserve both slave-holdings and financial resources. At the same time that it prosecuted the French and Dutch traders for denuding the Company's territories of slaves, it absolved both itself and specific indigenous regimes from the ambit of such legislation.[17] In 1795, five years after the Proclamation of 1789, the Governor-General in Council granted a *dastak* to a merchant called Muhammad Karbalai who along with other sundry articles imported nine Abyssinian slaves from Muscat as a 'gift' to Nawab Sarfaraz-ud-daula of Lucknow.[18] Nor, when it came to the Company-led armies in India was the Council averse to considering the indemnification of indigenous holders whose slaves

[15] Court's letter to President and Council at Fort St. George, July 1782 in Home Misc. 163, pp. 175–82, specially paras 16–17.

[16] Letter to Court of Directors, 17 January 1794, in *Fort William-India House Correspondence, 1792–95* (Delhi, 1955), XVI, p. 317. Approximately 1500 slaves were transferred annually from Bengal and Malabar to the Dutch Cape Colony between 1777 and 1801, *see* Shell, *Children of Bondage* p. 41, fig. 2.1.

[17] *See* for permission to procure slaves from Patna for the Awadh ruling house, Captain Gabriel Harper to James Alexander, 25 March, 1770, BS&MC, 15 April 1770, unnumbered.

[18] *CPC*, XI, p. 233, no. 914; p. 323, no. 1320 and p. 329, no. 1343. In 1796, the same merchant transported five Georgian slaves from Muscat to Calcutta, a'gift' sent by a resident of Baghdad for the Nawab Vazir at Lucknow. In 1833, another consignment of eighteen slaves, brought from Mokha by two merchants to Lucknow, were not sent back for reasons which were both financial (the British Government would have had to incur the expense of the return passage) and diplomatic. Besides,

served in these.[19] In 1799, two Malay male slaves were officially procured by the Company for the Botanical Garden in Calcutta.

It is clear that at least the late eighteenth century regulations were no part of the 'chain of events' leading to delegalization of slavery in 1843. The first link in the chain, depending upon the historian, has been 1774 (for Cassels) or Richardson's minute of 1808 (for Gyan Prakash).[20] Considering the brakes applied by the Company's judicial and executive officers to the abolition of slave-transfers and slave-holding within the territories of the Company, both the chain and its origins prove to be imaginary. The ways in which the Council in India sought to minimize the impact of Parliamentary Statutes, particularly 51 Geo III, Cap: 23 of 1807, remove further doubts on this score. Copies of the Statute were forwarded only to magistrates of the seaports, as the Governor General in Council 'did not consider its provisions to be applicable to the importation or removal of slaves by land'.[21] The construction that the Council at Fort William put on these Acts of Parliament were (a) that these acts had been confined to the transportation of African slaves only to various parts of His Majesty's or the East India Company's territories (b) the powers vested in the several Admiralty Courts were limited to offences of importation and removal of slaves by sea. Other kinds of transfers could only be tried in England. If the Acts had been intended to apply to inland transfers of slaves in the territories subject to the Company, 'every native carrying or removing a slave from one part of those territories

---

the newly styled 'King' of Awadh had offered to provide for them as 'servants to respectable people' which was gladly accepted by the Governor-General in Council. The Court of Directors clarified that there was to be no interference in the traffic so long as the slaves were not 'kidnapped' from the British territories. *See* correspondence in BC F/4/1479/58157.

[19] Sub Sec. Council Chamber to Sec. to BOR, 11 May, 1798, BBOR, 15 May 1798, no. 23. The instructions are to ascertain 'what consideration would satisfy' Raja Munnynath Singh whose slaves had joined the Ramgurh Battalion. For the original complaint *see* W.W. Hunter, *Bengal Ms. Records*, (London, 1894), III, no. 7404.

[20] For the full text of Richardson's proposals of 1808, inspired partly by the Parliamentary Act of 1807, Utilitarianism and Evangelical arguments, *see* BCrJC, 15 March 1816, no. 47.

[21] *See* Judcl. letter from Fort William to Court of Directors, 29 October 1817, paras 149–70, in Correspondence, E/4/98.

to another is liable to be sent to England, to be tried for felony'. The transportation of 'whole nations among whom domestic slavery had existed', it was urged, could surely not have been the object of Parliament.[22]

The care with which 'domestic slavery' (as a territorial phenomenon) was safeguarded by the Company is disclosed by Regulation X of 1811. Represented as the definitive attempt of the Company to prevent 'importation' of slaves from foreign territories, this Regulation entered a caveat for the trade of British subjects. Thus clause 5 of Regulation X said that for the more effectual prevention of the importation of slaves at the port of Calcutta, captains or supercargoes of vessels, *with the exception of the Honorable Company's ships importing at Calcutta,* should execute a bond of Rs 5000 with the Company before landing their cargoes.[23] Considering that the Act of 1807 had already prohibited all trading in slaves, this exemption not only contradicted Parliamentary statute, but significantly reinforced the monopolistic position of the Company *vis-a-vis* other maritime powers. The records of the Marine Department substantiate this interpretation: even in the 1830s stray commanders of the English East India Company's ships imported slaves from Africa to India.[24]

Islamic Law: Multiple Meanings and Singular Forms

Texts like the *Hidaya* and the *Sirajjiya*, compiled and published under the aegis of the Company after 1773, were meant to prove to legal opinion in England that slavery was enjoined by 'positive' law in India.

[22] Ibid. *See also* Judcl. letter from Fort William to Court of Directors, 30 January 1813, paras 146–55, in E/4/85. Both letters emphasized the fact that the statutes did not affect to the smallest degree the relation of master and slave wherever it had existed before the Acts.

[23] For the text of the regulation as prepared by Colebrooke, Lumsden and the Nizamat Adalat see BCrJC, 6 August 1811, no. 62. In this draft, the exemption of the Company's ship-masters from any bonds are repeated in two separate sections. For the final draft, *see Regulations Passed by the Governor General in Council of Bengal* (London, 1829), III, p. 460. Anderson, in 'Work Construed', argues that the scope of Regulation X was the trade in African slaves, but offers no explanation of clause 5.

[24] Letter of Marine Department to Court of Directors, 15 April 1831, BC F/4/1263/50837*. This refers to John Croft Hawkins, Commander of Company's sloop of war, *Clive*, who had carried 34 'boys' from Africa to India in 1830, and who were subsequently assigned to the Indian Navy.

Though the *Hidaya* was almost wholly about slave-related prescriptions, yet it was completely silent on the question of who was or was not a slave. Contrary to the silence of the texts, local practice defined the liabilities of both enslavement and transfers of slaves. Terms of enslavement and transfers were very clearly understood within indigenous adjudicating regimes. Not only was there an emphasis on documentation of transfers, (Table I, Appendix II) but the subject of the transfer was clearly described. Thus in a decision given on 25 February 1774, in a complaint filed by Naintarra against four men for unlawfully detaining and making her a slave, the maulvis of the Faujdari Adalat in Calcutta found that 'Naintarra was of good parents and never was a slave'.[25] The woman was declared a 'free woman' and the men were punished to varying degrees. The possibility that recourse to formal adjudicatory mechanisms in the eighteenth century occurred as acts of 'registration' in turn, is suggested by the number of decisions of this court that declared title to, or the jural status of, the slave. (Table II, Appendix II).

Those who *were not slaves,* whether as a result of a ransom agreement, or because of their 'good birth', had their status declared just as clearly as did those who were. Such declarations of status and title could be asked for in the name of both slave and master alike. (Table II, Appendix II). At the same time, the terms of transactions in slaves appear to have been protected by some guarantees, both of quality and in monetary terms. Thus in the instance of delivering a lame girl in exchange for Rs 28 advanced by the plaintiff, the girl was returned and the defendant was ordered to repay the sum he had received. Or in the case of non-fulfilment by the purchaser, of the terms upon which a transfer had occurred, the seller too had some guarantees. Thus when Gonga Bistno complained against Purbee for taking back a girl he had sold to the plaintiff, the Faujdari Adalat found that the defendant had sold his daughter to the plaintiff for Rs 20 ready and Rs 3 to be paid monthly. The plaintiff had apparently failed to pay the monthly sum. Repossession of the girl followed upon the monthly payment ordered by the Court.

[25] Proceedings of the Faujdari Adalat, 1774, P/154/39, case no. 333. The judges of this court were Qazi Abdullah, Darogha, Qazi Ghulam Imam, Ahsanulla Mufti, Nur-al-Huda Maulvi, and Mahmud Ruffadi Maulvi.

The last-named dispute suggested that a kind of conveyancing existed, which could not be comprehended within simple or finite categories like 'sale' 'hire' or 'pawn'. Each could slide into the other and the qualities of each transaction infected others. Slaves were also mortgaged against loans by primary masters, and the inability of the debtor to repay the creditor generally led to a transfer of title in the mortgaged or pawned slave. For instance, a letter from a Brahmin of Ratnagiri to one Gadgil suggested at least three different ways in which slaves could be offered against a loan. The first was a simple mortgage without possession, with his slave girl (*batik*) Mani pledged for a sum of Rs 100. If this was not acceptable, he offered to deliver possession of the slave-girl along with other goods worth Rs 75 in return for the money, and the third was an offer of outright sale of the girl for the same amount.[26] From Bihar and Bengal such transfers, by mortgage bond, of slaves, through the mid-nineteenth century[27] signalled a complexity in slave-transfers largely overlooked by historians.[28]

One of the ways in which slaves either fed themselves, or earned wealth for their holders, was through an agreement of 'farming' (ijāra) between holders and users: the holder received the wages the slave earned, or the money paid by the lessor. Such conveyancing was generally widespread in Bengal in the late eighteenth century, as well as later.[29] For instance, the *Seir* speaks of Banny Begam, third daughter of Rabia Begam, having,

[26] Rajwade (ed.) *Bharat Itihas Sanshodhaka Mandala Varshik Itivritta* (Shaka 1835/ 1923 CE, Pune 1923). pp. 191–2. For another transfer of a slave-girl by a khismatgar on account of a debt of Rs 30, *see SSRPD* III, p. 249, no. 1097.

[27] *See* Qeyamuddin Ahmed, 'A Nineteenth Century Case of a Long-Term Lease, Not Sale, Of Human Beings', *IHR*, 15, 1–2, 1988, pp. 276–80, for a bond which stipulates that the old owners 'agree not to bring any charge of abduction (against the lease-holder)'. Also *see* Chattopadhyay, *Slavery in the Bengal Presidency*, p. 15; for a bond by which a loan of Arcot Rs. 48 was raised against slave-collateral, *see* OIOC, Mss. Eur. F. 193/45, no. 6.

[28] The only discussion on slaves who were subjects of contracts and leases between holders has been by Sebastian Joseph 'Slave Labour of Malabar in the Colonial Context' in S. Bhattacharya (ed.), *Essays in Modern Indian Economic History* (Delhi, 1987), pp. 46–54.

[29] For descriptions of *ijāranamas* conveying slaves on long leases of 81 years, *see* Offg. Commr. of Circuit 10th. Divn., 24 July 1833, in *Constructions of Regulations and Acts Issued by the Court of Sudder Dewanny Adawlut from 1798 to 1847* (Calcutta, 1855), p. 295, no. 812.

formed into a band, some of her slave girls, joined to a number of other loose women, which she had taken into her pay, she got them instructed in the arts of dancing and singing,... commenced giving entertainments to Aaly-hibrahim-qhan... observing that one of her girls had made an impression on the Qhan's heart, offered her to him, adding that she was a girl of her's and that she had made him a present of her person....'[30]

Though this account is silent on any possible sums that may have been received by the mistress, other descriptions of such mechanisms were more forthcoming. Describing the 'naches' in the house of Raja Ramchunda during the Durga Puja of 1819, one correspondent pointed out that the performer named Bonnoo Jaun had recently been 'married for *three months only,* to a rich Mogul merchant, who paid One thousand Rupees in cash, as a Marriage settlement, besides Two Hundred Rupees to be paid monthly'.[31] Such conveyances of skilled slave-girls, for short or long tenures, and the non-payment of the stipulated sum by the hirer/holder to the mistress of the slave-girl led to litigation as in Mussamat Chutroo vs. Mussamat Jussa.[32] The mistress of Chutroo, the slave-girl, brought a suit for the recovery of Rs 1400, the arrears of a monthly payment of Rs 25 which the hirer, Babu Sarabjeet Singh, had apparently contracted to give the mistress. The provincial court of Benares had not only decreed this sum with arrears for four years in favour of the mistress, Jussa, it had ordered the payment of the stipulated monthly sum as long as Chutroo stayed out of the control of her mistress. Stripped of the confusing details, this was essentially the case of a mistress suing a man who had hired her slave and had failed to pay the monthly hire.

Though we know nothing of what Jussa in turn was meant to pay either to the state, or to anyone else, mistresses like Jussa were important to the economic and symbolic wealth of hegemonic revenue-collecting households of each region. An instance of the former were the sums raised by the zamindari of Nadia from the Kasbi-ka-Chauth, a 'Salamy paid by the keepers of Brothels on the admission of every new prostitute' which amounted to Rs 440 annually.[33] As

[30] Haji Mustafa (trans.) *Seir ul Mutaqherin*, III, pp. 85–6.

[31] S. Das, *Selections*, I, pp. 356–7.

[32] *Report of Cases Decided in the Sadr Diwani Adalat*, III, pp. 188–91.

[33] Enclosed in Collr. of Nuddea, Redfearn, to C. Stuart, 8 March 1791, BOR

with the Murshidabad household studied earlier, evidence from the trial of the Raja of Birbhum, Muhammad Zaman Khan, indicates that such slave-holding and conveyancing also overlapped with the constitution of kinship-economies. One of the main charges against the Raja had been that he had defaulted on his revenue-payments, and on the payments of his household because he had distributed a lakh of rupees 'amongst singers and dancers and in gifts to the servants'.[34] However one of these skilled women, and superintendent of seven slave-girls, was also the Raja's grandmother, 'his father having kept her Daughter and had a son by her—the Nautch set of Punna Boiee is now at Naghur, but not servants of the Rajahs'.[35] The *taifa* (literally group) of Punna Bai 'attended' upon the Raja periodically: either for this, or for the relationship in which Punna stood to the Raja, she received Rs 100 a month from the treasury.[36]

Multiple strands emerge—the connection between the acquisition of slave-girls by a mistress to the provision of skilled and unskilled services to a local hegemon; the complexity of the relationship between the mistress or superintendent of the slave-girls and the kinship structure of local households; the twin participation of the mistress both in the monetary economy of the market (wages) as well as in the gift economy of kinship (stipends); the simultaneity of transactions in two directions, one going from the Raja to the mistress and to her slave-girls, and another going from the mistress to the Raja. The most significant aspect of such multiple strands of conveyancing was the range of different economic actions (purchase, hire, lease or mortgage between different 'domestic' sites) that a slave might experience in one life-cycle. Such multiple transfers and positions could hardly be comprehended within the simple binaries of 'slave/kin' or 'domestic use/trade' or indeed even 'domestic/agrestic' upon which English legal thought was based. Given this complex form of conveyancing slaves, sometimes involving multiple parties, the non-fulfilment of any one

(Misc), 1 April 1791, unnumbered. In its response to the Collector, the Board of Revenue was completely silent on the Kasbi-ka-Chauth.

[34] Extract Proceedings of the Sadr Diwani Adalat, 6 July 1796, of J.H. Ernst, Actg. Collr. of Birbhum vs. Raja Mahomed Zeman Khan in BC F/4/19/767.

[35] Deposition of Mahomed Jaumil, 6 May 1795, ibid.

[36] Deposition of Dillore [Dilawar] Zemaun Khan, 22 May 1795, ibid.

party's commitments could bring upon it a range of charges, chiefly the charge of 'theft' of the slave.

As a culpable action, it had a long pedigree.[37] One of the legal forms in which disputes between owners and lessors, purchasers, and hirers appeared to have been lodged as judicial complaint was that of 'theft' which removed a slave from the jurisdiction of a previous owner/master. This form of complaint was not specific to indigenous regimes alone, but appears to have been shared by most of the eighteenth-century mercantile European powers.[38] Though such a charge implied that this action had robbed the owners of their 'property', behind it lay an anxiety that 'other slaves (would be) faithless to their masters and run away from them on knowing and hearing of the approach of the foreign vessels'. Such complaints thus simultaneously embodied the contradictory pressures of establishing the claims of masters in specific slaves, and denying the agency of slaves in forsaking them.

Both the form of such complaint, and its implicit contradictions, resonated in the practices of indigenous law-givers, who then reinterpreted older doctrine and texts into composite forms, as evinced in the Hidaya. Under the head of theft (*sariqa*), or the secret taking away of another's property, there were several actions and objects which did not merit the specific punishment (*hadūd*) fixed for this offence (amputation of the arm, in this case). So stealing a free-born infant, ('because a free person is not property, and the ornaments are only appendages'), or even an adult slave was barred from this punishment. The only slave-theft that deserved this punishment was that of the infant slave.[39] One way around this textual delimitation of actionable slave-theft may have been the representation of all slaves as 'children', encapsulated in words like *chhokra/chhokri* and *beta*/beti. However, the records of the Calcutta Faujdari Adalat of 1774 reveal that much greater complexity was possible in practice. (Tables I and II, Appendix II).

[37] See the description of a claim for repossession of a slave-girl brought by a Hindu scribe against a soldier in N. Manucci, *Storia do Mogor*, I, p. 195.

[38] For Portuguese complaints against Capt. Mercer, *see* Governor of Macao to Generale and Gentlemen of the Council, 22 November 1773, BPubC, 11 April, 1774, no. 5.

[39] *Hidaya*, II, book VIII, pp. 91–2.

Apart from the emphasis on documentation as evidence, a notable aspect of the seventeen cases of 'enticement' decided in the Faujdari Adalat of 1774 was the preponderance of female slaves as the objects of re-possession. It was also plain that charges of enticement were not made against 'strangers'. This was particularly important since slaves and servants of the same household could have been disciplined by bringing such a charge against one for enticing another. For instance, Faizunnisā Begam, of the family of the deceased Siraj-ud-daula, charged a young *sipahi* stationed at the zanana of her son Murād-ud-daula, of having 'seduced a 16 years unmarried daughter... from the house'.[40] In 1828, Burrattee, a female slave of the household of Fakrunnisa, was punished by her mistress, for having 'enticed away' two other slave girls of the household and selling them.[41] The critical question of whether such slave girls and 'daughters' were fleeing, though not directly addressed by such actions, is suggested again by the 1774 records. In at least two instances, the volition of the slave-girl, in taking flight, was acknowledged by the court, even though such an action was found punishable. Local ballads suggested that such women and girls fled with people (also servants and slaves) they wished to be with, and from their current or potential masters.[42] Another significant aspect of these decisions was the variety in the punishments ordered for what appeared to be the same action. And, finally, the charges appeared to have been to elicit a judicial declaration of title between two owners, a fact recognized in the decisions of the maulvis. In other words, the same form covered a range of different possibilities for action by masters and holders.

[40] *CPC*, X, January 24 1792, no. 71, pp. 20–1. This appears to have been the main theme of Bharatchandra's *Vidya Sundar*, satirising the loss of a female from the zanana of the Burdwan Raj. The suggestion of the poet that Sundar had been put up to it by other women of the household is repeated in other ballads from East Bengal, like that of Rajkanya Rupabati in Kshitish C. Moulik (ed.), *Pracina Purbabanga Gitika* (Calcutta, 1971), II, pp. 109–64. In the latter, the Rāja's *nafar* (second/third generation male slave) was asked by the principal consort of the Raja to elope with the 'daughter'.

[41] Petition of Fakhrunnisā Begam, 9 May 1828, BC F/4/981/27673.

[42] *See* the ballad of 'Monir Ojha-Manjur Ma' in Moulik *Prachin Purbabanga* II, pp. 387–411, also that of 'Amina Bibi o Nachhar Maloom pala', pp. 327–78. Both problematize the 'marriages' from which both women take flight. In the first, the girl, an 'orphan' is picked up by an elderly healer, reared to puberty, and 'married' by

When levied against trading partners, or other equal or higher-status individuals, the allegation of theft of a slave was to allege a serious breach of trust. Thus both Jagat Seth in 1839, and the Nazim of Murshidabad in 1838, had complaints lodged against them of a similar kind. In the words of the former, he had 'entertained' a dancing girl Luckee but her 'mother... presented a Durkhaust to the magistrate falsely asserting that I had kept her daughter in confinement'.[43] In the case of the latter, the Agent reported that the 'girl' in question (then about 18 or 19 years old) had been

'sold to His Highness for 5000 Rs and a monthly salary of 100 Rs and that Wuzzeerooddeen and Bunnee Begum [the complainants] would gladly compromise their claims if they could get half of the above monthly allowance made over to them'.[44]

Maratha records also reveal that such charges made by particular masters against third parties for the enticing of slaves could function as fiscal instruments, since the adjudicatory regimes granted specific compensation to masters by the person in whose custody the slave was found. Thus a soldier, Ali Ismail Piyada, who stole (married nika according to the records) the slave-girl of Mane Khan Bargir, of the government cavalry, was asked, not to return the slave but to make monetary compensation to the previous owner.[45] Such monetary payments, for eloping with another's slave, was not limited to soldiers alone. A nephew of a *kasar* (worker in brass), who had eloped with the slave-girl of another man also paid a fine of Rs 66.[46] The inference

him against her wishes. In the second, the girl is offered for by a man of the village who clearly indicates to the father that she will be kept like a slave (bandi), since he is already married to a 'respectable woman'.

[43] From Jagat Seth Govind Chund, recd. 26 March 1839, enclosed in *MNLR*, II, p. 393.

[44] AGG Caulfeild to H.T. Prinsep, Sec. to Govt., 26 April 1838, *MNLI*, II, p. 171. In reply, Prinsep agreed that Banni Khanum's complaint was 'open to the suspicion of being mere attempts to extort money', in letter to Caulfeild, 2 May 1838, *MNLR*, II, p. 331.

[45] V.T. Gune, *The Judicial System of the Marathas* (Poona, 1953), Appendix B IV, p. 362. The soldier was fined Rs 40, half of which went to the owner, and half to the state.

[46] Ibid., p. 363.

is that though 'slave-stealing' was recognized as culpable, it merited varying punishments ranging from paying compensation to the return of the slave, sometimes even a post-factum 'free gift' by the state. Thus the slave-girl who 'was corrupted' by a *gardi* (infantryman), Tom Ingrez, so that she ran away from her master, was given as *bakshish* to the soldier, even though he had been held responsible for her flight.[47] In a general sense, then, the allegation of a 'theft' of slaves occurred within the wider context of masters' ambitions and the judicial recognition of such claims and titles. But the actual repossession of the runaway/transferred slaves appears to have been left to the initiative of the proprietor, who could not automatically presume upon the complicity of the state officials in effecting such a recapture.[48]

As far as female slaves were concerned, the issue was complicated by a set of interlocking contexts: the first was that given the high demand for them in almost all sectors, both the leased-out female and the fugitive may have been further transferred by bonds of sale, mortgage, gift, or lease, depending upon the material resources of the household holding or acquiring her. The second problem arose from the nature of the pressures upon a female slave of reproductive age to take a 'husband' provided by her own master/mistress. This is suggested both by Haji Mustafa's account of his slave-girl married off against her wishes,[49] and by indigenous accounts. Some of the deeds by which masters of female slaves transferred them to other masters for marriage with the latter's male slaves in return for a fee indeed ended the jurisdiction of the female's former master: a deed dated 1805 transferring the slave-girl of one Brahmin to another Brahmin specified

[47] *SSRPD*, I, IX, p. 241.

[48] For an example of the failure to effect such recapture, *see* the decree of Sachala Misra, pandit in the court of the Raja of Tirhut, on the claim of Tularama Sarman for the recovery of his *ceti* Saito in K.P. Jayaswal, 'A Judgement of a Hindu Court in Sanskrit', *Journal of the Bihar and Orissa Research Society* (henceforth *JBORS*), 6, 1920, pp. 246–58; for a recent analysis of this document, *see* Richard W. Lariviere, 'A Sanskrit *Jayapatra* from 18th Century Mithila' in Lariviere (ed.), *Studies in Dharmashastra* (Calcutta, 1984), pp. 49–80; for a discussion of the same case, and of slave transfer deeds called *bahikhatas see* Upendra Thakur, 'Some Aspects of Slavery in Mithila in the 17th–19th Centuries', *Journal of the Bihar Research Society*, XLIV, I&II, 1958, pp. 47–52.

[49] Haji Mustafa, *Seir ul Mutaqherin*, I, Introduction, pp. 1–4.

that for a fee of three rupees, the power to sell or gift any children born of her marriage with the latter's male slave rested with the owner of the slave-husband.[50] Another deed from 1836 specified a small *munībāna* (master's fee) in addition to the price of the girl (sixteen sikka rupees) for the transfer of rights over her and her children born from the marriage with the male slave of the other master.[51] The range of measures that masters could resort to in arranging the connubium of their female slaves included the provision of a visiting husband, the *byakāra*, marriage to the male slave of the same master, and to masters' and mistresses' male kin and affines. Therefore, the location of female slaves within multiple and intersecting sites of conjugality also infiltrated the workings of the adjudicatory systems, where claims against specific people for having 'enticed' a 'wife' came to subsume the jural status (slave/ex-slave) of the woman concerned. Hence in the making of Company law from the eighteenth century, the pulls of different sites of 'domesticity' and conjugality in turn conditioned the remaking of slave-laws.

Remaking Domestic Mastery: Modification of Local Adjudicatory Norms

Diplomatic and fiscal concerns of the Company played a significant part in the passing of Regulations X of 1811 and III of 1832. Such enactments were invariably confirmations of actions and initiatives of local English magistrates, either refusing to, or in restoring fugitive slaves to their old masters. These were fashioned to *over-rule* specific indigenous norms and were based upon a partial misrepresentation of the transactions. Just as in the late eighteenth century so in the nineteenth, pronouncements of the Sadr Diwani and Nizamat Adalat were responses to local level magisterial intiatives. In 1798, the Judge of Chittagong reported that three slaves had run away from their Indian master; at least one of them was then 'servant' of Mr Coates, the Commercial Resident. The judge had suspended process 'for compelling his return', but asked the Nizamat Adalat for 'rules' to guide him in the matter. The latter responded by extending the scope

[50] Mitra *Early Bengali Prose*, p. 101, doc. 15.

[51] Ibid., p. 112, doc. 24.

of Regulation IV of 1793 (regarding the maintenance of the laws of the Muslims and the Hindus in suits of succession, inheritance, marriage, caste and religious usages) to slavery.[52]

Similarly, the initiative behind Regulation X of 1811 came from a case heard by the magistrate of zillah Gorakhpur between 30 March and 26 April 1810.[53] In this, Dusruth Tuppa [Thapa?] claimed three males and three female slaves who had 'absconded' from him to the lands of the Rajah of Butol. Dusruth, the Bakshi of the Nepal Durbar, demanded either that the slaves be restored to him or that he be recompensed in cash. The English magistrate asked for the opinion of the judges of the Sadr Diwani and Nizamat Adalat. In June 1810, the Register of the superior courts, J. Shakespear, wrote to the Secretary in the Judicial Department that though the rights in slaves were recognized by the laws in force, in this instance it involved surrendering the six slaves—'subjects of a foreign state'—or satisfying the monetary claim of the plaintiff. Instead of restoring the slaves to the complainant, on 6 July[54] the Governor-General in Council authorized the magistrate of Gorakhpur to pay Rs 226 to the claimant, but added

> Unqualified as the Hindoo and Mahomedan laws respecting domestic slavery at present are,... a regulation will be necessary, in order to establish the modification of it above noticed in the practice of our courts of judicature.[55]

The fact that the slaves claimed by the Nepali official were not returned in person but that a monetary compensation was ordered to be paid to the claimant was hardly the modification claimed. Indigenous adjudicatory bodies also granted compensation to masters in lieu of a restored slave. Before the Nizamat Adalat could prepare a draft of a regulation, Dusruth formally complained that 'children' from the hills

[52] Slavery in India: Correspondence (*PP*), 1828, pp. 74–5. This was the ruling that was upheld by the acting Governor-General in Council in April 1798. J.H. Harington the author of *Analysis of the Regulations and Laws of the Governor General in Council for the Presidency of Fort William*, was the Register of this Sadr Diwani Adalat.

[53] Slavery in India: Correspondence (*PP*), 1828, pp. 119–21.

[54] The date is wrongly transcribed in correspondence (*PP*), 1828, as 6 June, 1810. The comparison with the original BJC, (Civil), 6 July 1810, nos 2–3, reveals the error.

[55] Letter of G. Dowdeswell, Secretary to Government, Judcl. to Register of the Sadr Diwani Adalat, ibid.

were being conveyed into the Lower Provinces and sold there.[56] Thomas Brooke, the Agent of the Governor-General in the Ceded Provinces, issued orders to magistrates of Muradabad, Meerut, Saharanpur, and Bareli directing the restoration of what he imagined were 'fraudulently' acquired slaves.[57] In response, the Government informed him that the traffic

> not having been prohibited by a formal regulation of Government could not at the present moment be deemed absolutely illegal... and that the prohibition which the Agent directed the Magistrates to issue against the traffick in question... must be considered to be strictly speaking irregular.[58]

However, as the Government of the East India Company was at peace with Nepal, a 'foreign' country, and was trying to mediate a boundary settlement prior to the Anglo-Nepalese war, it was willing to consider the Bakshi's application seriously. Thus Regulation X was passed on 6 August 1811.

The 'moral principle, abstract justice and humane object' the Government thought was embodied in this regulation was actually the preservation of the rights of slave-owners in 'foreign countries' to recover the person of the fugitive instead of any monetary compensation. As conceded later, the Regulation of 1811 was to facilitate 'the restoration of the slaves of Governments of neighbouring states with which the Company's relations were friendly'.[59] Accordingly, section 4 of Regulation X stipulated that the magistrate shall 'cause to be sent back to their friends or country any slaves so imported'.[60] Implicit in this recognition of the rights of masters to have their slaves returned, was a belief that 'a slave by entering the Company's territories does

[56] Translation of letter from Buxee Dusruth Singh, 14 January 1811, in Extract Proceedings of Vice-President in Council, 26 April 1811, BCrJC, 6 August 1811, no. 60.

[57] For the priority of diplomatic concerns within which slave-restoration was ordered, *see* T. Brooke to Kajee Bimshaw, Soobah of Kumaoon, 22 March 1811, ibid.

[58] Extract Poll. Letter from Bengal, 27 July 1811, BC F/4/369/9221.

[59] Judcl. Letter from Bengal, 29 October 1817, BC F/4/1234/40338, and Correspondence, E/4/98.

[60] W. Blunt and H. Shakespear, *An Abstract of the Regulations enacted for the Administration of the Police and Criminal Justice in the Provinces of Bengal, Behar and Orissa* (Calcutta, 1824), p. 127.

not become free; nor can he who was lawfully a slave emancipate himself by running away from one country where slavery was lawful to another country where it is equally lawful'.[61]

The construction the Nizamat Adalat (there is no record of the muftis and qazi having been consulted when this was made) put on the regulation of 1811, narrowed down the field of its operation even further. According to the latter, the Act of 1807 prohibited only the importation of slaves 'for the purpose of being sold, given away, or otherwise disposed of'.[62] If Regulation X of 1811 was to be brought in line with the Parliamentary Act, it implied that even bringing 'foreign' slaves into the Company-held territories was permissible so long as there was no subsequent resale, especially a resale without the owner's permission.[63] John Adam, Governor-General till the arrival of Amherst, noted that Regulation X did not interdict 'the bringing in a man's domestic slaves for the purpose of keeping them'.[64] In short, the concerns of English jurists in the Sadr Diwani Adalat were limited to the rights of masters to move with their household-slaves between various territories. These concerns were written into the construction of Regulation X. Advising the magistrate of Agra in 1812, the Court observed that 'no part of the regulation in question was applicable to the sale of slaves not imported into the British territories'.[65] In other words, holding and dealing in 'indigenous' rather than 'foreign' (i.e. African) slaves was not criminal, nor were their transfers between different parts of the domain of the East India Company. Therefore, the Resident at Delhi, Charles Metcalfe, attempting to prohibit the resale-market in slaves, was instructed by John Adam in 1812 'that

[61] Judcl. Letter from Bengal, 1 March 1817, BC F/4/1234/40338.

[62] *Constructions by the Courts of the Sudder Dewanny and the Nizamut Adawlut of the Regulations and Laws for the Civil Government of the Whole of the Territories under the Presidency of Fort William in Bengal* (Calcutta, 1839–43), I, p. 26, no. 99.

[63] For a description of an African maid-servant, in the family of John Palmer, brought to India from England in the trousseau of Palmer's daughter, the wife of a British civil servant, *see* William Tayler, *Thirty Eight Years in India* (London, 1881), pp. 102, 112, 202.

[64] John Adam Papers, Mss. Eur. F.109/35, (Misc. Notes) folio 1.

[65] *Constructions by the Courts of Sudder Dewanny and Nizamut Adawlut* I, p. 26, no. 99.

persons already in a state of slavery were still to be liable to transfer by sale to other masters'.[66]

However, foreignness was a shifting criterion, sometimes defined in terms of territorial boundaries and sometimes of ethnicity. Thus when some police *darogahs* caught fourteen men from the territories of the 'foreign' state of Jaipur, who had brought fifty-nine predominantly female and young male slaves from Marwar to sell at Kanpur in 1813,[67] the Superintendent of Police of the Western Provinces urged the Nizamat Adalat to clarify that Regulation X was 'by no means intended... to prohibit or to interfere with the purchase or sale of slaves within the limits of the Company's territories, who may not have been so imported'.[68] The Nizamat Adalat obliged by circulating its construction of 1812 again to all the Courts of Circuit on 5 October, 1814.

Dissent and query from specific individuals like Richardson, the magistrate of Bundelkhand, or individual police officers or as in the case of Leycester, a judge of the Court of Circuit in Bareli,[69] revealed that the Company's government in India was not as monolithic nor homogenous in its policies as it would have liked to believe. Leycester, like Richardson, appeared to have distanced himself from the 'law', as practiced by the local English magistrates and that promulgated by the executive. In 1816, he had dissented from the magistrate of Farrukhabad who had ordered the recapture of a runaway female slave Gunna on the complaints of her mistress.[70] When Leycester was Chief Judge of the Nizamat Adalat between 1820 and 1823, he gave decision against another magistrate of Benares who had restored an identical slave fugitive to her mistress.[71] Yet in both these cases, Leycester's task

[66] Cited in E. Thompson *The Life of Charles, Lord Metcalfe* (London, 1937), p. 128.

[67] Magt., Zillah Kanpur, to Ch. Sec. to Govt., 19 July and 22 July, 1813, BCrJC, 7 August 1813, nos 42–3.

[68] For full text of the letter, *see* J. Carrau (ed.), *Circular Orders of the Court of Nizamat Adalut Communicated to the Criminal Authorities from 1796 to 1853* (Calcutta, 1855), no. 141, pp. 52–3.

[69] Slavery in India: Correspondence (*PP*), 1828, pp. 342–5.

[70] Report of the Law Commissioners, (*PP*), 1841, Appendix II, pp. 330–4.

[71] *See* Mussamut Chutroo vs. Mussamut Jussa in *Report of Cases in the Sudder Dewanny Adawlut,* III, pp. 188–91.

was simplified by the fact that the plaintiffs were 'prostitutes' and the defendants were important men of the locality. In presuming that 'free' people were kidnapped and enslaved, abolitionists like Leycester ignored the judicial evidence of the provincial courts, where cases of kidnap and theft were *mostly between slave-holders*, and the fact that most of the people said to have been kidnapped were those who were *already enslaved*.

These omissions were compensated by other British Circuit judges and magistrates. It was their complaints, rather than Leycester or Richardson's arguments, that were decisive in formulating legislation. For instance, a second judge of the Court of Circuit for the division of Dacca, R. Dick, partly on the basis of communication with the magistrate of Sylhet, had got to the heart of the matter in his report of 1813. In this, he reviled the traffic in slaves 'fraudulently' possessed by people who

> entice them to desert their masters, or, by the same seductive influence, cause them to be inveigled away through the medium of their private agents, and afterwards be sold at such distant places as to prevent discovery, or the return of the unfortunate being.[72]

Though persistently bracketed under the heading of the 'kidnapping of children' in the official records of the Court of Directors and of the Governor-General in Council, Dick's report minced no words on the need to restore these slaves to their 'families'.

When the Nizamat Adalat asked the Dacca Court of Circuit whether the persons described as 'inveigled' were free or slave, the former magistrate of Sylhet responded promptly. Most of the trials, he said, were for the offence of

> inveigling persons already in a state of slavery... the generality of the former description of persons above alluded to are females, who, being more employed for domestic purposes than those of the other sex, are in greater demand.... These females are carried to Dacca, Calcutta, Moorshedabad, Patna, and to those opulent cities which constantly insure for them a rapid and profitable sale.[73]

[72] Sir R.K.Dick to M.H. Turnbull, Register of Nizamat Adalat, 12 March 1813, BCrJC, 24 May 1816, no. 41.

[73] J. Hayes to H. Walters, Register to the Court of Circuit for Dacca Divn., 10 February 1816, ibid., enclosure in no. 42.

Other magistrates were equally emphatic. The assistant in charge of zillah Dacca Jellalpore referred to the common pattern of disposing of slaves by registered deeds of sale thus:

It rarely if ever happens that persons in a free state are inveigled away under false pretence... unless in the instance of young females who being obtained from their friends under pretence of marriage are disposed of either to public women or to rich individuals as servants.[74]

By 1 May, 1816, the Nizamat Adalat was convinced of the expediency of 'including some express provision' on the illegality of inveigling slaves, in the draft of a regulation required by government.[75] The main concern of the Bengal Government was 'that the future purchase or transfer of slaves shall be regularly registered'.[76] The great store set upon regulating slave-transfers through legal tender corresponded with the stamp paper of 8 annas value on which a master/lessor had to submit a claim in a slave according to Regulation I of 1814.[77] Besides, the emphasis on registration, contracts and bonds, appeared particularly ominous in the aftermath of the ordinance of 23 April 1816 passed by Moira's government. By this ordinance seamen (*lascars*) and menial servants alike were to be coerced into accepting contracts and into fulfilling them.[78]

This provided the background in which Richardson's proposals were finally discussed by a court that had effectively suppressed his minute of 1808.[79] For the moment, the Nizamat Adalat assured

[74] H.M. Pigou to H. Walters, 1 March 1816, ibid., no. 44. *See also* E.J. Harington, to H. Walters, 7 April 1816, for the opinion that no prosecutions had been occasioned by the enticing away of free persons in Mymensingh.

[75] Register of the Nizamat Adalat to Sec. to Govt., Judcl., 1 May 1816, ibid., no. 42.

[76] Sec. to Govt. to Register to Nizamat Adalat, 24 May 1816, ibid., no. 46.

[77] Cited in Draft Regulation submitted by J.H. Harington, 21 November 1818, BC F/4/1234/40338.

[78] 'A Rule, Ordinance and Regulation for the Good Order and Civil Government of the Settlement of Fort William', 23 March 1816, BCrJC, 26 April 1816, no. 27.

[79] Judcl. Letter from Bengal, 30 August 1827, BC F/4/1234/40338. An anonymous pamphlet later correctly pointed out that pro-slave 'suggestions were either allowed to slumber in silence in the offices of the supreme government, or were brought to light merely for the purpose of being formally condemned and abandoned'. *See* Anon., *Slavery and the Slave Trade in British India: with Notices of the Existence of these Evils in the Islands of Ceylon, Malacca and Penang* (London, 1841), p. 13.

Leycester that Regulation X would be extended to emancipate slaves 'imported' into British territories, except '*bona fide* domestic slaves, whom their owners might have possessed for one complete year before their importation'. The latter would be registered, as suggested by Government. As for the children of slaves and their liability to sale the Nizamat Adalat urged that 'whilst it is allowed to remain, with respect to the progeny of existing slaves, born under the British Government in the West Indies, and South Africa, the abolition of it... could not... be consistently proposed for India'.[80] The sum of these discussions was Regulation VII of 1819—the preamble spoke of the 'peace and happiness of families' which was destroyed by 'evil disposed women, who are employed to entice and take away the wives, or female children, of the fixed inhabitants from their respective houses, for the purpose of rendering them prostitutes, or concubines, or of otherwise unlawfully disposing them'.[81] The context within which this regulation was framed fleshes out the substance of the familial language, and its anti-slave direction. Clause II provided that 'enticing' any female—whether she was a 'married woman living under the protection of her husband, or of any person having the care of her in his behalf' or whether she was an 'unmarried female under the age of fifteen years, living with her parents or other legal guardians, or any persons acting in their behalf'—and disposing of her without the consent of the husband or parent or other guardian, was liable to imprisonment for six months.

As if to ensure that the language of kinship did not mislead the English magistrate, clauses V and VI empowered the magistrates to also proceed against those 'workmen and domestic servants' who 'wilfully quit the service engaged for before the expiration of the term agreed upon' (even though it acknowledged that many domestic servants did not enter into such 'engagements'). In the consolidated

[80] Resolution of the Court of Nizamat Adalat, 12 June 1816, ibid.

[81] The text of Regulation VII of 1819 is printed in *Papers relating to East India Affairs: Regulations passed between 1810 and 1824* (House of Commons, 1825). For the argument that this Regulation attempted to formalize the cohabitation arrangement of the lower classes while fixing financial responsibility, and was the particular contribution of the European household in India, *see* Singha, *A Despotism of Law,* fn. 18, pp. 125, 147, 150.

reproduction of Regulations of 1774 and 1816, such people who 'wilfully neglect to perform the work so contracted for' were made liable to imprisonment of one month for the first offence and longer terms of imprisonment for repeated misdemeanours of this kind. Only those recalcitrant 'workmen and servants' who could prove that their desertion was occasioned by 'gross maltreatment, or by non-payment of wages due' to the satisfaction of the magistrates could escape punishment.

This Regulation did not just tie *all* slaves down in whatever location, capacity or function they were found, it made the 'husband, parent, guardian' the sole and rightful authority for disposing of, and discharging, the slave in question. In other words, the right of the holder to initiate a transfer was secured unequivocally.[82] Furthermore, since, by law, only 'injury to property' justified its application, Regulation VII of 1819 could not be invoked if the child-woman had not been property in this sense.[83]

The Regulation then went on to throw the financial costs of such insurance on to the holders rather than on to the Company. In Clauses III and IV, it provided that all 'husbands' and 'fathers' (of 'legitimate' and 'illegitimate' children) were to maintain their 'families' in a suitable manner, as long as the woman did not find another 'protector'. In Clause VI section three a similar provision was made for the payment of wages to the discharged servant by the master.[84] Fully implemented, these would save the Company the financial burden of looking after decrepit unwanted slaves, and the fleeing and lost ones.

[82] The Magistrate of Murshidabad, A. Smelt, invoked Regulation VII against the eunuchs and servants of the de'orhi of Babu Begam for inveigling away the slave-concubine of Afzal Ali, a son of Nūrunnisā Begaṃ, *see* AGG Thoresby to Sec. Macnaghten, 28 November 1833, BC F/4/1522/60090A.

[83] Opinion of Register of the Nizamat Adalat, NWP, Agra, in letter to Register of Nizamat Adalat, Fort William, 8 October 1844, regarding Ramjeewun Pandey vs. Hingoo Pandey and Bishnu Pandey, BC F/4/2294/118156.

[84] By subsequent judicial interpretation, the ability of servants of long service to recover their wages shrunk. Under a construction of 1831, a servant could not sue for sums exceeding one year's wages, and by other constructions of 1821 and 1846, suits of servants against European/British masters under Regulation VII were disbarred from magistrate's courts. *See* Henry Carre Tucker, *My Notebook of Rules and Regulations, Collectory and Faujdary* (Calcutta, 1850), pp. 274, 389.

Conflicts in the Bureaucracy

Secular concerns determined just which master/mistress would succeed at a particular point in time in other regions.[85] As the company annexed new, hitherto 'foreign', territories, the contradictions within the administration of the Company increased. Magistrates, judges and revenue officers were caught between revenue collection from the newly-annexed region, (allowing the revenue-defaulting masters to sell their slaves and pay their dues), and simultaneously ensuring that the slaves thus sold/transferred/mortgaged could not be reclaimed. Legal guarantees protecting the creditor's right to sell the mortgaged slave of the debtor had to be balanced against fiscal claims made as compensation for a slave lost to the original holder or debtor.

The need to maintain the viability of such contracts and agreements *between* masters was optimised. Consider the case of Assam—the region the Governor-General in Council claimed had inspired Regulation III of 1832.[86] In 1823, before the Anglo-Burmese war broke out, the Commissioner of the Rangpur Division, D. Scott, like his counterparts had also restored runaway slaves to their 'foreign' (presumably Burmese) masters. However, the onset of the Anglo-Burmese war, and famine, made Scott go further. In 1825, he permitted male paiks whose labour had been the currency in which the Assamese kings had paid their ministerial officers, and who had been important in the war with Burma, to 'sell' and bond themselves to individuals and creditors.[87] Thereby, Scott eroded a subtle but important distinction between slaves and paiks: in this region, all male slaves were *exempt* from army-service, and were thus privileged *vis-a-vis* the paiks.[88] As Scott justified it in 1828, allowing the alienation

[85] For the restoration of eight runaway slaves of the widow of Nana Farnavis in 1816, despite the opinion of the Advocate General of Bombay that such restoration was contrary to 51 Geo III, c. 23, *see* Banaji *Slavery*, p. 315; for the arrest of three runaway African male slaves by the magistrate of North Konkan in 1818, on the complaint of their owner, Mrs Nesbit, a resident of Bombay, ibid., pp. 318–19.

[86] Judcl.letter from Bengal, 31 December 1832, paras 18–19, Correspondence on Slavery, *PP*, 1834–9, p. 77.

[87] David Scott, AGG on North East Frontier, to Ch. Sec. GOB, 4 March 1828, BC F/4/1115/29887.

[88] Barooah, *David Scott*, p. 163, fn. 18 and Amalendu Guha, 'The Medieval Economy of Assam', *Cambridge Economic History of India*, I, p. 503.

of 'public' labour to 'private' individuals went hand in hand with the resumption of the land that had constituted the paik's wages-in-land. By increasing the number of slaves who would not be liable to military service, Scott simultaneously demilitarized the Assamese monarchy and guaranteed that the new British-controlled Assamese state did not demand the services of the 'privately'-held slave. Such a guarantee, and the transfers, had been confirmed by the Government of Bengal in 1829.[89] Scott subsequently raised the matter of protecting the new slave-owners' and creditors' interests further.

The question rested on the judicial implementation of Regulation X of 1811 by a magistrate against a native of Patna for taking his Assamese slave with him out of the province.[90] Since Assam was no longer considered a 'foreign' country, the new owners, he inferred, should not be punished for removing their slaves to Bengal. Scott, who had pleaded that owners of slaves should be allowed to transfer slaves to meet their revenue payments, clearly saw Regulation X of 1811 as an obstacle in the path of the liquidation of slave-mortgages and debts. For him the basic issue, after 1826, was the protection of the creditor's rights to 'remove' or to sell the slaves of his defaulting debtor, without which the creditor himself may have been unable to meet the demands of the Company.[91] If the construction of 1812 was to hold, then transferring slaves between and within Company's territories, should also be applicable to Assam. Though Scott did not appear to have been

[89] Letter from Court of Directors to the Governor-General in Council at Fort William, Poll. Department, 10 March 1830, BC F/4/1371/54510. A contemporary Assamese account suggests that when the Burmese invaded Assam, they imposed a graded *kharikatana*, (poll tax) on all Assamese, including one of 8 annas on freedmen, bondsmen and slaves. Amongst the measures Scott instituted in Assam after 1826, was the confirmation and extension of the tax payable by slaves. *See* S.L. Baruah, 'Sadar Aminer Atmajivani', *BPP*, 97, 1, 1978, pp. 79–85.

[90] D. Scott to Sec. to Govt., 25 September 1830, BC F/4/1454/57705.

[91] D.Scott, AGG on North-East Frontier, to Ch. Sec., 25 March 1829, BPC, 10 April 1829, no. 60; Scott to Swinton, 31 December 1829, BPC, 26 February 1830, no. 17. In the second letter, Scott recommended that the government acquire the slaves of the debtor (and defaulter) by paying the creditor (with whom he has mortgaged or pawned his slave) a fixed rate for the slaves which the creditor is not being allowed to sell. This kind of sale alienation of mortgaged or farmed slaves was precisely what Reza Khan had tried to prevent, in redraft of Cornwallis's *ajir* proposals, *CPC*, VIII, nos 1325–6, and no. 1379, pp. 585–88.

aware of the Construction, or of the Circular Order of 1814, this was broadly the permission he asked the Bengal government for.

Led by the reference to 1811, the latter thought that the issue was the affirmation of rights of all Assamese masters in their slaves, and the latter's immobilization within Assam. Accordingly, it asked the Nizamat Adalat to formulate a new regulation to prevent the 'importation' of slaves from Assam into other British-held territories. Both the central executive and the judiciary confirmed the rights of Assamese masters to have their slaves restored to them in the event of flight or kidnapping. The draft regulation of the Nizamat Adalat stipulated that

> All slaves removed by sea or by land, without the permission of the Governor-General, from provinces which were not dependent on the Presidency of Fort William at the date of the enactment of Regulation X, 1811, [i.e. Assam] into the provinces which were, would be considered free; and all persons concerned in these transfers would be liable to six months' imprisonment and fine not exceeding 200 Rupees.[92]

Bentinck's council insisted on its modification, on the grounds that it distinguished between the owners bringing slaves from newly acquired territories into Bengal, and owners living in, and transferring slaves between, the old provinces. While Regulation X had only penalized 'removal of slaves for the purpose of traffic', the proposed enactment threatened to end their removal at all.[93] Hence, Regulation III of 1832, entitled 'A Regulation for extending the provisions of Regulation X, 1811', was anti-slave.[94] It ensured slave-owners of Assam that they

[92] Register of the Nizamat Adalat to J. Thomason, Dy. Sec. to Govt., Judcl., 19 August 1831, BCrJC, 8 November 1831, no. 47.

[93] W.H. Macnaghten, Sec. to Gov.-Gen. to J. Thomason, 30 December 1831, BCrJC, 13 March 1832, no. 21, W.H. Macnaghten to J. Thomason, 26 October 1831, BCrJC, 8 November 1831, no. 50.

[94] For the full text of Regulation III, *see* Papers Relating to Slavery in India (*PP*), 1834–39, pp. 394–95. By specifying that 'all slaves... removed... *for purposes of traffic...* are hereby declared free', it left the traditional opt-out clause for slave-holders, that slaves brought from Assam (or elsewhere in India) were for domestic purposes, not resale. For the opinion of English judges of the Bombay Faujdari Adalat that neither the exportation of a slave and his subsequent sale in a 'foreign' territory, nor the selling of a 'free child' into slavery, was punishable by Regulation XIV of 1827 of the Bombay Code, *see* NAI, Legislative, 12 October 1840, nos 37–8.

were protected against the 'removal' of their slaves by others not authorized to do so. Clause III specified that persons concerned in the sale and purchase of slaves 'so removed, knowing him or her to have been so removed' on conviction before a magistrate would be liable to imprisonment.

Since the restoration of fugitive slaves was viewed as an important part of diplomacy by the Governor-General in Council, the problem that a local administrator like Scott had referred to, could not be allowed to resolve itself by default. For instance, in 1836, when the Political Agent at Dehradun refused to restore the fugitive slaves of the Raja of Garhwal, he was reprimanded by his superior, the Lt. Gov. of the North-West Provinces for 'having exceeded his powers'.[95] However, the eleven fugitive slaves demanded by Nepal Darbar were tacitly allowed to remain in Tirhut.[96]

The differential priorities of local and central administrations revealed that the central government in this period itself gave contradictory assessments of Regulation III. In 1833, the Commissioner of Circuit for Assam submitted a list of rules to be established by proclamation which included provisions for the slave-born children in Assam to be bound 'to serve their Parents or owners' until the attainment of eighteen years; the registration of slaves and slave-born with patwaris of villages; the rights of 'parents' to sell their 'children' in times of distress, and so on.[97] T.C. Robertson who had served as Commissioner of Assam before being appointed to the Sadr Diwani Adalat, urged that the practice of borrowing money by pledging

[95] Correspondence Regarding Slavery (*PP*), 1834–39, pp. 58–9. For intra-administrative conflict around the ransoming of approximately 10,000 slaves from their Khamti and Singhpo masters in Upper Assam, in order to form them into a militia under British command, *see* F. Jenkins, AGG North East Frontier to W. H. Macnaghten, Sec. GOB, 16 January 1836, and Major White to F. Jenkins, 15 December 1835, BPC, 9 February 1836, no. 2 and no. 5; for Jenkins' interpretation of Regulation X of 1811 permitting settler Singhpo masters to claim their runaway slaves within British jurisdiction, *see* IPP, 7 March 1840, no. 172.

[96] Resdt. Nepal Residency to Magt. Tirhut, 13 April 1836 and correspondence, NAI, FPP, 2 May 1836, nos 66–8.

[97] F. Jenkins to R.D. Mangles, Sec. to GOB, Judcl. Deptt., 22 August 1833, and Correspondence, Report of the Law Commissioners, (*PP*), 1841, Appendix VI, pp. 405–9.

'personal service', was very common in Assam and any rule declaring invalid a suit for reclaiming the services of a bondsman would abrogate the practice. He was assured by government that no such consequence was intended by these measures, that such personal service should not be transferred by sale to a third party by the creditor, nor the slave/bondsman transferred out of the jurisdiction of the erstwhile owner.[98]

To other presidencies, confused by the contradiction between the Supreme Government's permission of internal transfers in 1811, and the prohibition of such transfers in 1832, the central government painted Regulation III in different colours. The government of Fort St. George pointed out to the Bengal government that clauses I and II contravened its own construction of the Parliamentary Act of 1807, by now appearing to disallow the transportation of slaves from the erstwhile 'foreign' territories to the Company's jurisdiction.[99] In response, the latter pointed out that no such contradiction existed: Sec. II, it said, 'does not declare the removal of slaves by sea to be punishable', but that slaves removed after 1811 were to be considered 'free'. Moreover, the penalties provided for in the Regulation were not attached to the 'removal' of slaves but 'for subsequent sale or purchase of slaves so removed'.[100]

Strictly drawn notions of territoriality, of the finite boundaries of each economic transaction, and the bifurcation of 'domestic use' from 'economic exchange' had evidently come to bear on distinguishing 'removal' and 'trade' in slaves. Thus even with the 'foreign' African slaves imported between 1833–45, the fractures between central and local administrators persisted. In 1834, the Chief Magistrate of Calcutta, D. McFarlan, asked the Bengal Government to enable pilots of vessels coming 'from a certain quarter of the world' (presumably the Gulf) to report to the police every case where they might suspect that slaves were on board.[101] The Marine Board, and the Government

[98] Judcl. Letter from the Governor of Bengal, 6 July 1837 in Papers Relating to Slavery in India (*PP*), 1834–39, 51, no. 697, pp. 79–80.

[99] Sec. Fort St. George to Ch. Sec. to GOB, 7 September 1832, BCrJC, 25 September 1832, no. 8.

[100] Sec. to Govt. to Actg. Sec. to Fort St. George, 25 September 1832, ibid., no. 9.

[101] D. McFarlan to Sec. to Marine Board, 26 May 1834, BC F/4/1528/60416.

of Bengal, while broadly accepting such a suggestion, demurred that the pilots should not be held responsible 'should any slaves be landed from any particular vessel'.[102] So when one young male slave, employed as a lascar by his owner, Ali Abdulla, in the *Aden Merchant*, was suspected of having been sold at Calcutta in 1843, the Chief Magistrate of Calcutta wrote:

The greater portion of the Arab vessels coming into Port have slaves either as seamen or domestic servants—So long as trafficking with them is not practised within these Territories, the Police have not considered themselves bound to interfere.[103]

Official mythology that slave-use in a 'domestic capacity' was separable from 'sale' ignored the complex cycles of 'use' and 'exchange' within which a slave-lifetime could be spent. Nusseed, the African slave who had originally belonged to Ali Hamed of Aden, had been sold to Ali Abdulla. Then, together with three other male slaves, he had been put to work as a lascar in the ship. Ali Abdulla had charged the owners of the ship, Sorabji & Co. for their wages, as an earlier master too had done when putting Nusseed to work as a *khalasi* in a boat.[104] The domestic authority of the master or mistress determined whether he/she would be the subject of other economic transactions. Ajeema, the ten or eleven year old female slave was kept for three or four years in her master's house in Farrukhabad before being taken to Muscat and exchanged for a Swahili male slave.[105]

[102] Sec. to Marine Board to D. McFarlan, 31 May 1834, and Sec. to Govt., Gen. Dept., to D. McFarlan, 9 June 1834, ibid.

[103] Ch. Magt. of Calcutta to Sec. to GOB, Judcl., 10/20 March 1845 (error in original letter), enclosure II in Sec. to GOI to Ch. Sec. to GOBy., 18 April 1845, BC F/4/2112/9946.

[104] Ch. Magt. of Calcutta to Sec. to GOB, Judcl., 20 November 1843, and Poll. Agent at Aden to Ch. Sec. to GOBy., 23 January 1844, BC F/4/2066/94848. The presence of such lascars was noted again later, cf. Ch. Magt. Calcutta to Sec. GOB, 23 March 1855, NAI, FPP, 19 October 1855, nos 18–19.

[105] Enclosure in letter from Secy. to GOBy., to Secy. to Government of NWP, Agra, 27 November 1843, in B.C.F/4/2112/99472. For an exceptional official criticism of abolishing 'trade' while maintaining 'domestic' use, *See* H.A. Churchill to Actg. Ch. Sec. GOBy, 22 January 1869, NAI FPP, A, July 1869, nos 230 and 233.

Rhetoric of 'Preserved' Law: Anti-Slave Justice

Judges of the Sadr Nizamat Adalat described their own proposals in the 1830s as attempts to ensure that the 'existing state of the law was altered as little as possible'.[106] In a similar vein, a private letter written by W.H. Macnaghten to Lord Auckland in 1837, warned the Governor-General not to introduce any 'rash measure' of emancipation since the practices of the courts were already sufficiently liberatory.[107] Referring to a Circular Order of the Nizamat Adalat which instructed local magistrates to refer persons claiming property in a runaway slave to the civil court to prove his 'right',[108] Macnaghten disingeniously asked, '...but whom is he to sue? He might as well sue an ox or an ass as the slave himself and by suing him he ipso facto admits the invalidity of his claim'. In the event of the slave being him/herself sued in a civil action, the courts would uphold the 'law of the defendant' and this would prevent 'a Mahommedan from owning a Hindoo slave and vice versa'.

Was Macnaghten right in claiming that civil suits for repossession of slaves offered relief to slaves? Regulation XV of 1805 had provided for the appointment of 'Mahomedan and Hindoo Law Officers' of the Zillah and City Courts to the posts of *sadr amins* in the zillahs and cities. These maulvis and pundits had been allowed, as sadr amins, to try referred cases to the value of Rs 100. From Bakarganj, it was reported that the maulvi in his capacity of sadr amin in 1820–21, had thrown out two claims in slaves as fraudulent.[109] In Chittagong, another such sadr amin had instructed a 'Mussulman native of some respectability' applying for possession of the 'daughter of a poor woman that had been sold to him by the mother' to recover the purchase money from the mother.[110] Another principal sadr amin in Tirhut

[106] J. Thomason to W.H. Macnaghten, 8 Nov. 1831, BC F/4/1454/57705.

[107] W.H. Macnaghten to Lord Auckland, 8 April 1837 in British Library, Broughton Papers, Add. Mss. 36473, folios 105–110.

[108] For text of two constructions one of May 1830 and another of March 1835 to this effect, *see Constructions by the Courts of Sudder Dewanny and Nizamut Adawlut of Regulations*, I (Calcutta, 1840), no. 550, p. 232; II (Calcutta, 1839), no. 939, p. 205. [For an opinion on the legal propriety of such a suit, *see* Macnaghten to Maj. White, 12 September 1836, NAI, FPP, 12 September 1836, no. 103.]

[109] Report of the Law Commissioners (*PP*), 1841, Appendix II, p. 287.

[110] Ibid., p. 289.

suggested that Regulation III of 1832 annulled all transfers and sales of slaves in the region.[111] However they increasingly found themselves at odds both with the Sadr Diwani Adalat as well as with provincial English officials. For instance, the sadr amin of Tirhut found his interpretation of Regulation III struck down by the Magistrate and Judge of Tirhut and the Sadr Diwani Adalat. Between them, the latter agreed that 'Regulation III of 1832 does not prohibit the transfer of slaves for money; it merely prohibits the removal of them for the purpose of traffic from one territory, British and foreign, to any other territory dependent on this presidency'.

Macnaghten's implication that slaves themselves could not be sued was belied by the number of cases (at least ten were reported) that the Sadr Diwani Adalat had decided between the period 1830–40 in which slaves themselves had been made defendants in civil suits for repossession. Furthermore, in at least two instances, the dispute was originally between members of the same 'family' over the inheritance of rights over slaves, one party accusing the other of having inveigled the slaves.[112] As Saiyid Afzul Ali illustrated when he used Regulation VII to effect the recovery of his concubine in Murshidabad in 1833, indigenous claimants appeared to have gradually acclimatized to these changed norms in bringing claims in slaves. Thus he brought a regular suit against 'the slave, a girl and Salamut Ali, the person who was harbouring her; he got a decree and the girl was restored to him'.[113] Another man spoke of his father who had bought Chanda and Nida from their master: the women 'refused to come under his dominion, whereupon he sued them in the civil court of Sylhet... I continued it, and got judgement, which was confirmed on appeal in the provincial court of Dacca'.[114] The civil and sessions judge of Sylhet in 1836 had confirmed the fact that 'on a suit instituted in the civil court, and the claim being proved, an award is given declarative of the master's right

[111] Judge of Tirhut to Register of the Presidency Court of Sadr Diwani Adalat, 20 March 1835, in *Constructions*, II, p. 214, no. 955.

[112] J.C.C. Sutherland (ed.), *Select Cases in the Sudder Dewany Adawlut*, V, p. 71 and VI, p. 52. Other instances of intra-family disputes over the control of slaves can be seen in the complaint of the Diwan Deo of Kuch Bihar against his paternal cousin, the Raja of Kuch Bihar, in Extract BRC, 22 August 1812, BC F/4/442/10641.

[113] Report of the Law Commissioners (*PP*), 1841, Appendix I, p. 227.

[114] Ibid., p. 256.

to the slave: and on application from the master, the slave is apprehended and delivered to him. If the slave refuse to serve or to comply with the award, he is imprisoned so long as the master chooses to pay the subsistence money'.[115] In these disputes, decrees of 'costs' against an appellant or respondent could have either left the slaves in the possession of whoever had them, or induced their further transfer and capitalization; judgements rarely gave emancipation to the slaves in question.

Despite half-truths and outright lies, Macnaghten's letter clarified that the recovery of a slave by a civil action meant that the slave had to be categorized as property, especially after Regulation VII of 1832 stipulated that all rights of property were liable to the jurisdiction of civil courts.[116] Once this was done, the rules applicable to property became not 'Islamic' or 'Hindu' law, but those derived from colonial slave-law. The Register of the Sadr Diwani Adalat instructed the Officiating Judge of Zillah Shahabad in 1836 that suits to establish rights in slaves 'must be looked upon in the same light as other personal property'.[117] In addition it was considered 'highly inexpedient that such cases should go before a native should the reference of them to a European judge be practicable'.

The inference that the replaced indigenous adjudicatory norms and standards of proof by more emphatically anti-slave laws is borne out by the evolution of criminal regulations as well. Most indigenous adjudicatory regimes had distinguished between the body of the slave and the free, as they distinguished between the body of one's own slave and that of another's slave. These distinctions were eroded from the onset of Regulation law. Commenting on the report of cases of 1773 of the newly constituted Nizamat Adalat, Hastings overturned a judgement of *diyat* given by the maulvis against a man who had beaten his slave-girl to death on the grounds that '... as it does not appear that there was any intention of murder and Mohammedan

[115] Ibid., Appendix II, p. 301.

[116] When asked for opinions on the rights of masters to sue for their slaves in court, officials cited Regulation VII of 1832 more frequently than either Regulation III of 1832 or Regulation X of 1811; hence many English officials believed that rights in slaves-as-property were liable to litigation and decree in civil courts.

[117] *Constructions*, II, no. 1022, p. 264.

Law *as well as ours* admits of moderate correction to a slave or even a hired servant, the prisoner ought to be acquitted or forgiven'. The principles of English slave-law, in the image of which 'indigenous' law was to be made, was evident when Hastings transformed another sentence of diyat against a man who had killed another's slave-woman into one of death for the offender.[118] The uniformity of the maulvis' decisions—granting diyat against both men—contrasted sharply with Hastings' differentiating between the two, acquittal for the former and death for the latter. Contemporary Maratha records suggest that the maulvis at the end of the eighteenth century had been closer to other indigenous legal decisions on such issues. For instance, Malhar Balaji Mahajan, who beat his slave woman to death had to pay a *gunehgari* (fine) of Rs 25, and perform a *prayaschitta* (expiation for sins).[119] Instances of monetary compensation for injury to a (presumably third party's) slave had persisted in many regions. For instance, in 1791, the Qazi and Mufti of the Patna Court of Circuit found four men guilty of having wounded a slave and ordered that compensation be made.[120] In contrast, Hastings' decrees, while allowing the encroachment on another's property to be visited with capital punishment, also accepted that the master could destroy his own slave.

In 1796, for instance, when an English indigo-planter Mr. Hunter, was tried in the Supreme Court for having severely wounded two slave-girls of his own household, the counsel for the defence, Srettell, contended that 'slavery was legal by Mahomedan law', to argue that Hunter should not be punished.[121] The much-vaunted Regulation VIII of 1799 withdrew the exemption of the master from *kissas* (claims of compensation in blood or money from the kin group of, or the individual, who had taken the life of a slave) but paid little attention to whether it was a third party's slave that had been destroyed or one's

[118] *See* Niharkana Majumdar, *Justice and Police in Bengal: A Study of the Nizamut in Decline* (Calcutta, 1960), Appendix A, pp. 315–16.

[119] V.T. Gune, *The Judicial System of the Marathas* (Pune, 1953), Appendix B4, p. 354, no. 56.

[120] K.P. Mitra, 'Matters Criminal (from Behar Records)', *BPP*, 73, 1954, pp. 127–36.

[121] For depositions in the Magistrate's court at Tirhut, *see* BCrJC, 14 October 1796, no. 19.

own. The abrogation of masterly exemption (for murder of slaves) appears to have established certain levels of physical brutality as 'minimum' and the yardstick of 'moderate chastisement'.

Procedural changes under the aegis of the Company consolidated the marginalization of certain norms governing slave-delicts. By Regulation I of 1810, judges of the Court of Circuit, which had supervisory functions over magistrates' courts, had been allowed to dispense with the fatwa in criminal cases wherever they deemed it advisable. Regulation XVII of 1817 provided that wherever the judge of a Court of Circuit differed in his opinion from the fatwa of the Qazi or other 'Mahomedan law-officers', the proceedings were to be submitted to the Nizamat Adalat. If two or more judges of the latter court disagreed with the fatwa they could overrule it, and pass sentence accordingly. A minor example of the reversals that could follow upon such procedural changes should suffice. In 1820, a burglary was committed in the house of an eunuch, Amanee. Asked whether he suspected anyone; Amanee named Pimmee who had demanded that Amanee send his chela Khudabaksh to him (for implied sexual purposes), but had been spurned. The Magistrate of Bareilli caused all three to be apprehended. The fatwa of the 'law officer' of the Court of Circuit declared that Amanee was liable to a lesser punishment: the Judge of the Court of Circuit differed on the grounds that Amanee had brought up the chela, possessed a certain degree of authority over him and should be equally punished. In the Nizamat Adalat, the man Pimmee received the smallest punishment, and the slave Khudabaksh the highest (thirty stripes with a rattan and imprisonment with hard labour for eight years).[122] Such judgements actually reversed the exemptions of slaves from lesser punishment reserved for slave-delicts in sharia law. In the Quran every punishment specified for a free man or woman was halved when it was a slave who was the guilty party. While indigenous jurists tried to absolve slaves from *tazeer* on the grounds of 'minority',[123] the direction of change in the colonial courts

[122] Government vs. Pimmee and 2 Others, 3 October 1820, *Report of the Nizamut Adalut*, (henceforth *RNA*), II, p. 49.

[123] *See* case no. 660, Mr. Thomason against Mohammud Ally and Phini Slave Girl, in P/154/39; for similar absolution of a 12-year-old girl for cutting off the penis of the man she had been sold to, see Juttee Ram vs. Jye Munnee, RNA, Vol. II, pp. 29–31.

appeared to have been towards removing such absolutions. Regulations that disbarred justificatory pleas of adultery as grounds of murder, as did Regulation IV of 1822, actually appear to have weighed heavily against those husbands of female slaves who resisted the predation of other males on their slave-mates.[124] Such male slave-spouses actually bore the brunt of regulations like IV of 1822, addressed to cases in which the fatwa absolved the guilty slave from punishment. Similarly, when a female Dher ran away from her master and refused to return, muftis of the Faujdari Adalat in Madras exculpated the fugitive. Reiterating the principle that only captives in war and their descendants were considered 'true' slaves in 'Islamic' law, they issued a fatwa which by virtue of this strict delimitation of slave-status, absolved those 'received from their parents during famine or at other times' from the punishment reserved for flight.[125] This pro-slave interpretation was objected to not only by the English judge of the Madras court, but also by Amos, one of the Law Commissioners, who later characterized this fatwa as 'gross misrepresentation'.[126]

The Courts and the Protection of the Domestic

As Auckland had admitted, the law 'depends upon the opinions of those by whom it is administered and is liable to fluctuate with a change of functionaries'.[127] The consolidation and continuity of official insistence on the rights of masters, in an apparent closing of ranks against the abolitionists in the metropolis[128] as well as against ameliorationists in the Company's administration, and against the opinions of maulvis favouring a pro-slave ethic in judicial cases, should further qualify any consideration of the claims of delegalization of slavery by Act V of 1843. To contemporaries, especially slave-holders,

[124] Russoo vs. Mohun Launda, RNA, II, p. 67.

[125] Fatwa of Muhammadan Law Officers, NAI, Legislative, 6 September 1841, nos 3–16, ff. 11, 81–2.

[126] Minute by A. Amos, 5 August 1841, BC F/4/1947/84542, ff. 132–5.

[127] Minute of 6 May 1839, British Library, Auckland Minute Books, Add. Mss. 37711, folio 60b.

[128] *See* H.T. Prinsep's Minute of 31 July 1841, scathing about the 'exaggerated feeling' aroused by the abolitionist movement in England and Europe, in British Library, Broughton Papers, Add. Mss. 36473, folios 66–9.

this Act offered a glimpse of apocalypse. From Assam, 'almost all the higher classes' protested against the promulgation of this Act.[129] One group of land-holders from Sylhet predicted that it would 'tend to the ruin of all India, especially that of the respectable part of the population'.[130] These holders knew that the object of the act 'is not emancipation of slaves' but the issue of restoration of fugitive slaves ('forsaking their services') by the Diwani or Faujdari courts—the sum of Section II of Act V. Again referring to this section of the Act, another group of slave-holders from Bakarganj remonstrated that this would take away the 'control of masters.... Because if the rights of persons in slaves cannot be enforced in courts the term "slave" will be obliterated'.[131]

The fears of indigenous slave-holders were hardly justified. Though the Governor-General had claimed that 'Magistrates do not interfere for the restoration of a runaway slave to his Employer',[132] officers in Bengal had certainly not earned such pronouncements of faith. Few magistrates doubted the 'legality' of their own actions in catching and restoring runaways. For instance in 1844, the Magistrate of Murshidabad 'arrested four children... who state that they have escaped from the Begum Ruhunissah's [sic] Dheoree' and asked the AGG to ascertain whether anything had been stolen. In response, Sitanath Bose, the Diwan of the Nizamat, confirmed that 'ten maid-servants of the Begam' had absconded, four of whom were attached to the

[129] F. Jenkins, Commr. of Assam, to Under Sec. to GOB Judcl., 9 May 1844, BC F/4/2107/98688. The fear of aristocratic displeasure, and demands of compensation, also ensured that Act V of 1843 was not promulgated in any of the 'native states' evident in a list in BC F/4/2261/114628. For official justification of non-promulgation in these states or 'foreign territory', *see* BC F/4/2233/111702. For the consequent prolongation of the trade in African and indigenous slaves, *see* NAI, FPP, A., February 1862, nos 51–3, October 1862, nos 7–8, November 1862, nos 111–13, May 1865, nos 89–91, April 1874, nos 137–58, July 1883, nos 14–25. For persistence of magisterial action regarding fugitive female slaves, *see* BC F/4/2112/99462 and BC F/4/2112/99475 and BC F/4/2112/99476.

[130] Abstract Translation of a Bengali Petition from the Zamindars, *Mirasdars, Taluqdars, Pottahdars* and Ijaradars of Sylhet, (680 signatures), 23 February 1843, ILP, 7 April 1843, no. 13.

[131] Translation of a Bengali Petition from the Landholders and others of the Dist. of Bakarganj (125 signatures), 10 March 1843, ibid.

[132] Minute of May 6, 1841, in British Library, Auckland Papers, Add. Mss. 37713, folio 102a.

Toshakkhāna, two to the *Rakābkhana*, and four were personal attendants.[133] Similarly, when certain inhabitants of a *mohalla* in Hughli town, who were 'not relations of the children' claimed that boys between the ages of six and seventeen had been 'stolen',[134] magistrates promptly swung into action.

If these functionaries did not slacken in their pro-mastery ethic, nor did the judiciary. The ethic of 'equality' of slaves and masters in law was guaranteed by a clause of Act V, that a wrong inflicted upon a slave was as amenable to redress as a wrong done to a free man. Yet, while criminal courts were allowed to prosecute eunuch slaves of the Nizamat in 1853 for the murder of a slave-servant,[135] the Abyssinian-born eunuch Nazar Ali Khan could get no justice on the basis of a similar equality before law. He charged the AGG, Torrens, with forcibly placing a guard over him and his property. In the *rūbakari* held by the Officiating Magistrate, a letter was read from the Advocate General which said that for purposes of Act 53, Geo.3, c.155, the prosecutor had to prove himself to be a native of India, which 'Asiatics (sic)... are not'. Nazar Ali was accordingly called and asked to state from what country he came, 'but he was unable to name the place of his birth, or that of his Father-....'.[136] The case against Torrens collapsed.

Another aspect of delegalization was that no rights arising out of an alleged property in the persons and services of a slave would be enforced by any civil or criminal court or Magistrate. The records of both provincial Sessions and Presidency Sadr courts in the 1850s reveal that slave-holders continued to bring private prosecutions for their runaways and conveyed slaves, and the courts continued to declare entitlement to them. In the case of Musst. Kumollee vs. Amjad,[137]

[133] Magt. Murshidabad to AGG Berhampore, 11 August 1844, and Sitanath Bose to AGG, 11 August 1844, *MNLR*, II, p. 502.

[134] Magt. Hughli to Under Sec. to GOB, 24 June 1845, BCrJC, 9 July 1845, no. 236.

[135] Upon the Sadr Nizamat Adalat finding the evidence insufficient to hold Aman Ali Khan and the eunuchs guilty of privity, the Governor-General took extra-legal measures to punish both Aman Ali Khan and his young master, *see* Sec. GOB to AGG, 15 November 1853 and note by Dalhousie of 9 November 1853, NAI, FPP, 26 May 1854, nos 72–3.

[136] Enclosed in Collr., Bhagulpur, to AGG, 2 February 1852 in *MNLI*, II, p. 677.

[137] *RNA*, II, pt. II, 1852, p. 281.

the sessions judge of Tipperah, H.C.Metcalfe, recorded that a pregnant 'female of very low standard of intellect' had been induced to accompany a man Tumeezoodeen to his house by a promise of marriage. In association with Amjad, he sold the woman for the sum of Rs 8 to one Arif Chowdhury. The latter, 'happening to hear that a chuprassee... had a claim upon the woman, became alarmed and after fifteen days, returned her to the sellers'. The second time around, they offered her to the wife of one Ahmed Meeah for Rs 6, but the latter declined to make the purchase in the absence of her husband. The matter eventually reached the magistrate. The Sessions Judge had noted that this was a woman 'of mature age, perhaps of 25 years, and fitted and intended for domestic service'. Yet because he thought a crime of slave-stealing had been committed, he recommended that Amjad be given two years' imprisonment with labour.

Upon this the Nizamat Adalat, with R. H. Mytton as the Officiating judge, decided that though the selling of 'free-born persons into slavery is an offence by the Mahomedan law' this principle did not apply in this case. As magistrate of Sylhet, Mytton had recorded his opinion that Regulation III of 1832 did not stop masters selling slaves from Sylhet in adjoining territories. As a judge in the 1850s, he noted that the woman calls herself the chhokri of Zukky Chuprasee: 'the usual term in the eastern districts for a female slave'. This being the case, the issue that had to be decided was whether the selling of a slave by anyone *not her owner* was an offence. According to his interpretation of the concluding section of Act V of 1843, it was so.[138] What Mytton was looking for was fraudulent transfer: he found that no force, misrepresentation or deceit had been resorted to. The offer for sale was 'open', and the woman had made 'no effort to free herself, or object'. To the second sale she 'appeared to be an assenting party'. The man was released. Nothing is said about the female slave.

Since chhokri did in fact denote the female slave, it is clear that many cases involving such people were decided in the provincial and Nizamat Adalat, not in favour of the release of the girls and women believed to have been stolen, but in favour of one or the other claimant.

[138] Contemporary barristers like W. Morgan and A.G. Macpherson appeared to have interpreted Act V in an identical fashion in their *The Indian Penal Code (Act XLV of 1860)*, (Calcutta, 1861), pp. 319–20.

Thus in the case of Bindoo Bystumee vs. Ram Bagdee, Korebun Ali Sheikh, Puddo Khangy, Shama Khangy[139] an eight/nine year old slave-girl called Kamini Chookree but described as 'daughter' of Bindoo Bystumee, was believed to have been carried away by the two men and two women prosecuted. As in many previous and subsequent cases, the men and women accused of the theft were neighbours, people whom the child knew well enough to 'come to their house to play'. The accused pleaded that they had taken her away 'at her own request', but were disbelieved by the Sessions Judge of 24-Pergunnahs, E.Bentall. As he reported it to the higher court, the grounds for his disbelief were firstly, the fact that Shama and Puddo were 'prostitutes' and the 'abduction of children by such people is by no means uncommon'. Thus he had sentenced each to three years' imprisonment without irons and a range of fines. The higher court upheld the decision. No one commented on how Bindoo Bystumee had come by this chhokri. At the core of these cases was not just a conflict about the re-possession of the slaves, but the 'lawful authority' to transfer the slave-girl. As Mytton had indicated, most judges held that only the owner had 'lawful authority', and not third parties. Tests of 'lawful authority' to transfer a slave suppressed the question of freedom for such slaves, illustrated in Government and Bhoobun vs. Chundoo.[140]

The divergence of opinion articulated in administrative discussions resonated in judicial decisions regarding the 'purpose' for which a girl was sold. Thus if a slave was bought for 'marriage' or domestic use, it was not culpable, but if the slave was bought by a professional (*peshagar)*, the latter could be convicted. In Government and Musst. Kooroani vs. Adoo, Aradhun, Musst. Gourmonee Peshagur[141] the sessions judge of Bakargunj, C.Steer, found that Suffer Jan, a 'delicate looking thing of 9 years of age' was married to Aradhun, but directly after the ceremony, she was taken to Adoo's house. Adoo then took her to the house of the nazir of the Sadr Amin of Dacca, Rujub Ally. She fell ill there after four days, and was removed back to Adoo's house where she stayed nearly two months. At the end of this time, Adoo and Aradhun sold her 'under the name of Mina' to Gourmonee

[139] *RNA*, 1852, II, pt I, p. 837.
[140] *RNA*, 1855, V, pt II, p. 35.
[141] *RNA*, 1853, III, pt I, p. 630. Emphasis added.

for Rs 18. At both the thanah and the sessions court they admitted the last transaction, but offered what seemed to be important insights into just how this system worked. Aradhun claimed that Rujub Ally had given him Rs 15 to get a 'wife', on Adoo's security. Adoo also stated that he had stood security for the Rs 15 given by the nazir to Aradhun 'to procure a female slave', whose sickness caused Rujub Ally to threaten that 'if some better person than Suffer Jan was not soon procured, he would disgrace Adoo's female relatives'. Thus in consultation with Ram Nidhee Sarkar, a witness for the prosecution, Adoo sold her to Gourmonee for the stipulated sum, and returned Rs 15 to the mother of Rujub Ally, and paid a commission of Rs 3 to the Sarkar.

Critical to the decision in this case was the silence of the documents on the relationship between Aradhun and Rujub Ally, the role of securities in what was otherwise a 'benign' deed (getting married), the form of the 'marriage', and, most important of all, the complete absolution of Rujub Ally from the procedure of court. Thus even though the sessions judge admitted that 'the marriage of Aradhun was *a mere device to obtain a decent girl* for Rujub Ally', punishment stayed focussed on those who had already been vulnerable to this man. Despite evidence of this kind, where becoming one man's 'wife' also meant being the slave of the husband's patron-master, the courts erected a false dichotomy between immoral transfer (prostitution) and moral use (marriage).

The implication that professional senior slaves and concubines appeared to have been the final recipients in the long chain of slave-transfers had evidently aided many sessions judges, and the higher court, in determining the guilt of those who were accused of having 'stolen' and sold the slave-girls. Like Leycester, many overlooked the anti-slave direction of these judgements. Moreover, such indictments of the peshagar women were not indictments of prostitution *per se,* not at least, where the Nizamat Adalat was concerned. Thus in Government and Musst. Pakeeya alias Solabebee vs. Sheikh Shetabdee, Kurkomaree Peshagur, Assanoolla Chowkeedar and Sheikh Allum,[142] the Sessions Judge of Dacca had reported that Pakeeya, a ten-year-old girl had been brought from an indigo factory in Faridpur 'by the

[142] *RNA*, 1853, III, pt II, p. 643.

direction of a person in charge of it (whose name is very indistinctly given)' to the village of Kurrea, where she had been sold to Kurkomaree. The chowkidar and Sheikh Alum had acquired the stamp paper on which the endorsement of a bond for Rs 21 was signed 'written out as if from the child' (as if they were 'voluntary' bonds against loans of money). The Sessions Judge had convicted them all, but had pointed out that Kurkomaree 'had at one time been evidently the victim of similar villainy'.

Upon this the Judge presiding over the Nizamat Adalat, A.J.M. Mills, noted that the 'child' had not been taken by force or fraud 'out of the legal custody of its parent or guardian'. Indeed, the 'child' had been 'made over to him [Shetabdee] by another to sell for that person's benefit'. The Regulations, both the accused and the judge averred, did not provide against prostitution. This kind of perspective was one that was shared by administrators like Prinsep who in 1842 had announced that procuring of girls for prostitution was not a capital offence. He wrote,'... We are *not legislating to prevent prostitution* but to prevent the unfortunate victims being considered and treated as slaves'.[143]

It appears that authorities in England were misinformed about the import of the decisions in the Nizamat Adalat. In 1859, the, Secretary of State for India referred to these decisions above in reprimanding the acquittal of a woman in Monghyr, who had taken on a 'lease of ninety years' some slave girls referred to as 'infants'.[144] Stanley pointed out that B.J.Colvin and A. Sconce, two judges of the appellate court, had given 'legal sanction to a slavery of the most revolting and degrading kind'. The Government of India was unbowed. It referred the matter to the Legislative Council for consideration with the imminent Penal Code.

The Penal Code of 1861 formally declared the holding and trading in slaves criminal, but continued to punish, under section 361, the taking away of such girls (henceforth called minors) from 'lawful

[143] Minute recorded by H.T. Prinsep, 18 January 1842, BC F/4/1947/84545. Emphasis added.

[144] Judicial Despatch to India, no. 23, 14 June 1859, in British Library, House of Lords Sessional Papers, 1860, Second Session, VII; Judcl. Narrative of GOB, 26 February 1859, NAI, Home, Judcl., 25 November 1859, no. 14.

guardianship': the latter, of course, included, the 'master or mistress in whose service the child had been placed'.[145] The Code did not substantially improve upon the model prepared by 1838. Under the offence of kidnapping, for instance, there had been only two sections: the first was kidnapping from the territories of the East India Company, and the second kidnapping from lawful guardianship.[146] Lawful authority, enshrined in metaphors of kinship and eldership and the clauses of the Penal Code, was invoked time and again. In 1862, certain political officers in the Chittagong Hill Tracts brought to the attention of the Government of Bengal the existence of hire-mortgage bonds by the terms of which

> C being indebted to D, either verbally or by writing, binds his wife, son, or daughter... to serve D for a certain number of years, or C. sells his children outright.[147]

The Commissioner, C.T. Buckland remembered that even in 1845, creditors had regularly applied to the British courts for the enforcement of these bonds, without which they found regular payment of revenue impossible: hence he advised his junior officer not to interfere with the privilege of the 'head of the family' to hire out the services of any of his wives or children as he saw fit. The Government of Bengal agreed: a 'father can make contract of service for his child or bind him apprentice for a number of years' especially if the wife/son/daughter was under the age of sixteen.[148]

Soon after the Penal Code had declared the holding, buying and selling of minors illegal, the Government of India promulgated legislative measures that consolidated these transfers, and enlarged the domestic markets. Act XXII of 1864 and Act XIV of 1868—together referred to as the Contagious Diseases Acts—were passed to register all prostitutes living in cantonments and cities. Such registration comprehensively infringed the provisions of the Penal

[145] John D. Mayne, *The Criminal Law of India* (Madras, 1896), p. 638.

[146] *The Indian Penal Code as Originally framed in 1837* (Reprint, Madras, 1888), p. 63, sections 353–7.

[147] Superintendent of Hill Tribes to Offg. Commr. Chittagong Divn., 13 May 1862, WBSA, Judcl., December 1862, enclosure no. 142.

[148] Under-Sec. to GOB to Offg. Commr. Chittagong Divn., 11 December 1862, ibid.,no. 143.

Code the Government itself had contrived, especially Sections 372–3 which were supposed to prosecute the sale, purchase, disposal and use of 'minor' females for the purposes of prostitution. Evidence about many of these minors suggested that they were slaves. In 1864, a Deputy Commisioner of Police in Calcutta listed a number of judicial cases where nine to twelve-year-old girls had been sold and bought for the brothels in Chandni Chowk area in north Calcutta.[149]

After the famine of 1866–7, prompted partly by the decision of the High Court acquitting three elderly women of the charge of buying slave-girls,[150] the Government of India instituted an enquiry into the numbers of girls under ten years of age who were thus circumstanced.[151] The aims of such an investigation were clear: it 'would at most merely provide that such girls already so circumstanced and below the age of ten years, should be registered, without interfering in any way with their custody'. Neither the central nor the provincial administration thought that the 'possession' of such girls could be prohibited; nor was it necessary to pass any public proclamation against such transfers and possession.[152] Arguing that under Sec. 373, 'mere possession with criminal intent' was differentiated from 'obtaining possession with criminal intent', (a legal rephrasing of the domestic use versus resale argument), the Government of India advised the Government of Bengal against making the former culpable since 'we become involved in many very delicate and complicated questions and contingencies'.[153]

The language of familialism and familial authority upon which the courts proceeded reinforced both the languages of the indigenous

[149] Memo. by Dy. Commr. Police, Calcutta, n.d., WBSA, Gen. (Gen.), May 1865, no. 15.

[150] In re. Tofa Bai, Gunga Bai, Poonee Bai, Registrar of High Court of Judicature Fort William to Sec. of GOB, 15 June 1867, BJC, June 1867, nos 208–10. This case also began with the flight of the slave-girl Kunda, and a battle for repossession, *see* Bengal Police First Report, no. 14, dt. 25 February 1867, in Letter from DSP to Magt. of Cuttack, 26 July 1867, BJC, November 1867, enclosure, no. 123.

[151] Sec. to GOI, Home, to Offg. Sec. to GOB, Judcl., 9 April 1872, BJC, April 1872, nos 68–9.

[152] Sec. GOB to Sec. GOI, Home, 8 May 1865, WBSA, Gen. (Gen.), May 1865, no. 21.

[153] Sec. to GOI, Home, to Sec. to GOB, Judcl., 4 July, 1873, BJC, July 1873, no. 97; also Offg. Sec. to GOI to Sec.to NWP, 29 October 1878, NAI,Home Judcl., October 1878, no. 268.

holder and the predilections of a plantocratic English officialdom; it also tied together the two halves of the nineteenth century. Where a magistrate in Sylhet had restored the slave-born slave-'daughter' to the claimant 'father'in the 1830s,[154] magistrates in the seventies noted the languages of kinship set up by the buyer.[155] An Indian civil surgeon in Rangpur described the manner in which girls sold to elderly prostitutes, 'generally look up to the woman who brings them up as their own mother'.[156] The most comprehensive discussion of purchased kinship and prostitution came from another Indian deputy magistrate who wrote

> every old prostitute has her *sukri* and the wealthy ones have 3 or 4. As in the case of adoption by childless hindoos they take care with a view to the success of the ingrafting process to obtain children too young to know their parents, so the old prostitutes use a similar discrimination in the selection of their girls. Their ages vary from 1 to 5 years, they are taught to call them Mamma.[157]

In turn, such metaphorical sons and daughters were bequeathed to their patrons by dying 'mothers'.[158] It was this group of 'minors' who continued to be registered as 'prostitutes' in many parts of the city and province. Referring to the 'exceedingly common' practice in Mymensingh, of 'little girls... sold chiefly by "buragies" [bairagis] to prostitutes who regularly train them up to their future trade' another

[154] Report of the Law Commissioners (*PP*), 1841, Appendix II, p. 304.

[155] Dy. Magt. Nattore to Commr. Rajshahi Divn., 28 May 1872, WBSA Judcl. October 1872, no. B268. In Nattore town, he noted, of a total of nineteen boys and forty-one girls under ten years of age in the brothels, one boy and ten girls 'appear to have been obtained as gift from outside'.

[156] Civil Surgeon Rangpur to Commr. Rajshahi Divn., 27 May 1872, ibid., no. B269.

[157] Dy.Magt. Brahmanberia to Magt. Tipperah, 25 May 1872, ibid., no. B271.

[158] Will of Srimottee Wodoycomaree Bewah, L/AG/34/29/69, (1844), pt, I, pp. 369–78. This specifies that the patrons have direction of her *palak kanya*, translated by the court as foster daughter. Another *palito kanya* is left in the will of Sonah Bewah Bariwalli to the charge of her patron's heirs, L/AG/34/29/129, (1884), pt II, pp. 106–8. This kind of rearing of children does not appear to have been limited to girls, since the will of Rajcomaree Raur speaks of the fostering of a *bhikkaputro*, L/AG/34/29/96, (1857), pt IV, pp. 7–14, as does the will of Rajcoomaree Bewah with regard to a *dharmaputtra* reared in lieu of an 'womb-born son' in L/AG/34/29/129, pt II, pp. 25–8. The significant differences between the fates of such sons and daughters however awaits further investigation.

official spoke of bonds that were executed between sellers and purchasers to guard against 'the girl's friends inveigling her away after the sale had been completed' and the registration of these bonds in the local zemindar's *kacheri*.[159] The courts actually blessed such registration of transfers in a landmark judgement.[160] In an identical manner, magistrates restored 'wives' under the provisions of the Penal Code (497–8) in the eastern districts of Dacca, Mymensingh, Bakarganj. Individual local officials both in the 1830s and in the 1870s realized that terms denoting marriage overlapped where female slaves were concerned, yet institutionally little changed. To cite a laudatory government resolution, 'substantial justice is in many cases done by absconding wives being made over to their husbands even when no case lies against any abductor'.[161]

To urge then that the possession, transfers, and the use, of slaves between different domestic sites was delegalized by mid-nineteenth century is indicative of historiographical complacency. While earlier adjudicatory regimes had tried to settle matters regarding slaves through a finely calibrated set of distinctions, the Company's Regulations obliterated various gradations within slaves, and between slaves and non-slaves. As evident in Assam under David Scott, these actually led to the full legalization of slave-transfers in the name of a *lex loci* that was neither Sharia-derived Law, nor pro-slave, and the power of masters and mistresses was regirded by law as patriarchal privilege. Specific enactments of the Company which some historians have hitherto understood as abolitionist were in effect measures that consolidated masterly control and restored fugitives. The Company's readiness to reinforce the claims of masters, by special attention to complaints of enticement and theft, conditioned the enactment and construction of statute regarding the 'trade' in slaves. Thus the apparent delegalization of slave-holding by Act V of 1843 was undercut by the practice of the courts that continued to adjudicate in matters

[159] Offg. Commr. Rajshahi Divn. to ec. GOB, 21 November 1864, WBSA, Gen.(Gen.) May 1865, no. 18.

[160] Queen vs. Noor Jan and Juggut Tara, *Sutherland Weekly Reporter*, XIV, 1870, Criminal Rulings, Appellate, pp. 39–41.

[161] Magt. Bakarganj, cited in Annual Crime Report of Dacca Division for 1871, BJC, October 1872, no. 104.

concerning claims in slaves, and vitiated administrative practice into the 1870s. As an anonymous pamphlet had summed it up:

> whilst the utmost latitude of exposition has been given to every rule which sanctioned or was supposed to sanction the hateful system... the most limited interpretation has been put upon every law, British or native, the tendency of which was to curb it, or to destroy its revolting adjunct, the slave trade.[162]

In a curiously-timed permission that availed nothing, in 1876 Parliamentary Statute (39 & 40 Victoria) authorized the amendment of three relevant sections of the Indian Penal Code, 367, 370 and 371 to prevent further transfers and holding of slaves,[163] just as efforts were being stepped up to bring more slave 'minors' and prostitutes on to the official register in India.

[162] Anon. *Slavery and the Slave Trade in British India*, p. 46.

[163] *A Collection of Statutes Relating to India* (Government of India, Legislative Department, Calcutta, 1881), II, pp. 1015–18.

Conclusion

'It is an abuse of language to talk of slavery'.[1]

This study has highlighted the processes by which a key political and social institution was erased from the historical record. Partly a result of linguistic strategies within every group of slave-holders, the obliteration was also the result of contemporary, and subsequent, assumptions about the role of slaves in a political economy. While British officialdom occluded non-plantation models of slave-holding and use in order to maintain the socio-political institutions in which they were deeply imbricated, historians failed to discern the shifting nature of slavery in South Asia due to an absorption with fixed statuses and dyadic models of slavery. In order to correct this, I have provided a 'thick description' of different kinds of slaves as they moved *within* as well as *between* different households, within the Nizamat of Murshidabad, and similar households in Dacca and elsewhere.

This study has argued that colonial slave-formations and practices lay at the heart of ideologies by which 'elite' classes could be separated from non-elites, and genealogical kin could be separated from non-kin and 'illegitimate' kin. Standards of genealogical legitimacy led to the bowdlerised 'Hindu' and 'Islamic' law which assumed substantive and reified proportions in the courts. By a) privileging marriage, of whatever provenance, as the origin of progeny b) inserting all women and children into 'caste'-ranked systems of marriage and birth c) inferring social rank from the ritual undergone in the course of the

[1] A.Amos, Minute of 10 June 1842, BC F/4/1947/84545.

incorporation of the woman, and d) classicizing various local terms to fit them into the corpus of 'law', the Company's officers and agents bequeathed a new legacy to Indian social history. This legacy of thought bypassed the *actual* modes of incorporating slaves into kin-structures, especially as mothers. The ends of the reproductive labour of such purchased and kinless females lay not just in the lineage-members she biologically produced, but also in the hierarchies of deference, of eldership and juniority, of cousinhood and conflict that she reproduced, without which intra-household relationships would have collapsed. As with so many slave-mothers, the direction of their children's lives lay not with them but with older women who were de'orhi heads, or with the men who had hired or purchased them. The intra-generational conflicts that occurred as a result of inter-generational assimilation revealed both the assimilative and fissiparous nature of such households. Under Company rule, various epistemological, social and legal consequences followed, one of which was the establishment of systems of status-reckoning different from earlier ones. These changes eroded the privileges and possibilities of negotiation not just for the slave-born, but also for their mothers and fathers, within the groups of kin. The reconstitution of consanguinous kinship and the affective family began in earnest when the models of the devoted and kinless wife, or the kinless *gharjamai* dependent upon the household, were redeployed in rhetoric and reformist agendas of the later nineteenth century. By the end of the nineteenth century, a *gestalt* switch in the epistemological field became obvious: where kinship had been fraught, tense and metaphorical, and slavery real, now kinship was material and affective, and slavery alone metaphorical. In the process, slavery and gender became interchangeable signs. An acknowledgement of the specific and significant contribution that slaves had made to the physical reproduction of wealth and power, as well as to the lineage, family and community structures of the region became impossible.

Even as it imposed a standardized way of reckoning legitimacy in Anglo-Islamic and Anglo-Hindu law within which slave-born and slaves had earlier been articulated as claimants in the household, the Company backed off in implementing these doctrines uniformly or simultaneously. Even though patterns of slave-incorporation, concubinage and slave-birth were practically universal, the Company's decision in each case was made not in conformity with its construction of 'Islamic' law but according to exigencies of political control and revenue-collection at specific moments in each region. This can be seen at two different levels. In overtly Shia Muslim households like that of Murshidabad, Chitpur and Dacca till the 1840s, while the patterns of the haram continued to be similar, the criteria for the Company's recognition of particular males as 'heirs' to either the zamindari or the masnad fluctuated dramatically. Thus while the son of a slave-performer was recognized as Nazim in Murshidabad in 1836, and a slave-born son again recognized as the Nawab of Murshidabad in 1884, the son of a reared slave-girl was not recognized as Nawab of Dacca in 1843. Similarly, while the Company persistently claimed that the adjudication of inheritance issues was to occur according to principles of Islamic law, its officials adopted different strategies available to them. Together with the resumption of lakhiraj lands, and the resumption of stipends declared hereditary in the 1770s, the legal contortions deprived both the less wealthy slaves and erstwhile masters of material incomes, which were then re-circulated within the colonial structure. These changes then meant a rough equalization between slave, ex-slave and free in terms of wealth-holding over time and the erosion of slave-specific exemptions from punishment in law. Contrary to assertions about the ways in which textualization and codification imposed a standardized 'Islamic law' in the Company controlled areas, what we then have is a particularization and segregation of households from each other—the 'aristocratic' from the 'common', Shia from Sunni, Nizamat from Mysore Princes, Awadh from Bengal.

Within the overtly Hindu households, omitted from this study for lack of space, a similar pattern of fluctuations between colonial pragmatism and between what was thought to be doctrinaire law was visible, as in the case of Bharatpur early in the century. Elsewhere, too, though judicial decrees and tests of legitimacy were invoked, extra-judicial influence was important in households with a history of litigation around slave-born claimants. For example, in Tripura when the disputed succession to the Raj affected (through the desertion of the countryside) the collection of revenue, and a survey officer reported the obstructionism of those who supported the claims of the elder brother Neelkishen,[2] the Government of Bengal determined to bolster the contending *jubaraj* Beerchandra on the throne of Tripura well before the court's decision in the matter. The provincial administration directed the magistrate to warn all claimants to the Raj that any attempt on their part to weaken the authority of the Raja by exciting disaffection would be viewed 'with extreme displeasure'.[3] Similarly, in Keonjhur after the High Court had dismissed the Rani's claims on behalf of her adopted son, the Government of Bengal, threatened by imminent insurrection, in turn, dismissed the sirdars who acted as her emissaries in 'going from village to village, poisoning the minds of the people' against the successful Dhunonjoy.[4] The increasing role of extra-judicial management of issues arising from slave-birth made the personal opinions and predilections of local English officers far more important than judicial decrees, which were in many instances based upon information supplied by such officers. Simultaneously, where the Government of Bengal had taken over the management of these states, it was unwilling to increase the monthly provisions that had to be disbursed to these potential claimants and contenders within each household. In such cases, as in Kuch Bihar, it devised ways to control their futures through other administrative means. Reminiscent of the attempts to control collateral males, and concubine-born sons, in Murshidabad through the establishment of the Nizamat College, such schemes were reiterated for the management of the numerous males in Kuch Bihar, Tripura and the native states of Orissa. Dividing the group of collaterals in Kuch Bihar into legitimate and illegitimate—an official advised their segregation thus:

> Some of the illegitimate boys are sons of women of very low position indeed, and their numbers are such that if they were to be specially recognized when they grow up in consequence of their descent, the task of looking after and following them up in life would be a difficult and onerous one. It is very desirable that when they grow up they should merge into the general population of the State... the illegitimate boys... must go into the world and

[2] Rev. Surveyor 5th Divn. to Commr. Chittagong Divn., 16 January 1864, WBSA, Judcl. (Judcl.) October 1864, enclosure no. 46.

[3] Sec. to GOB to Commr. Chittagong Divn., 12 July 1864, BPP, March 1865, no. 12.

[4] Memorandum from T.E. Ravenshaw, 22 January 1866, and Offg. Suptdt. to Sec. GOB, 21 March 1866, BPP, April 1866, nos 28–9.

make their own living, in an humble position if necessary... In short, once grown up and well educated by the State they must shift for themselves, and their descent will not be recognized by the State any more.[5]

These schemes of segregation only confirmed the fault lines that already set one male against another. What was different was that where earlier some of these brothers would have lived like sons of the Raja ('given horses and ponies and everything else as freely as they are given to the Raja himself'), henceforth any material investment in them by the State would have to be paid for with ignoble rank and undignified treatment.[6] On the other hand, a parallel group, the Thakoors in Tripura, who all claimed descent from a former Raja, were differently deployed: there they continued to hold magisterial and judicial posts in the state and subdivisional administration without any fixed salary but obtaining either the 'necessaries for living' or 'large zamindaris' from the Rajbari.[7] This deployment of such collateral males constituted a saving in administrative costs for the Government of Bengal: the opposition of the Thakoors to the abolition of slavery was a small price to pay.[8]

However, just as extra-judicial management of households became more emphatic by the end of the nineteenth century, the judiciary's insistence upon 'proofs' of marriage also varied according to the kind of stakes involved. Where 'power' was largely made up of symbolic, non-monetized, honorific privileges, the courts were willing to admit certain social and ritual claims of the slave-born. Thus, at the end of the century, when the son of the Raja of Chhedra claimed the succession to the Rajgi, gaddi, and heirship of moveable and immoveable property and privileges by virtue of being son by a

[5] Commr. of Rajshahi and Kuch Bihar Div. to Sec. GOB, 18 April 1877, BPP, July 1877, no. 40.

[6] The Commissioner threatened one particular set of such sons that if they did not leave Kuch Bihar, they would be put into their *palkīs* by constables and kept there by force, ibid., no. 40. For details of the Jenkins' and Boarding Schools at Kuch Bihar where many of the 'illegitimate' Rajguns were educated, *see* Deputy Commr. of Kuch Bihar to Commr., 29 October 1877, BPP, December 1877, no. 10.

[7] Babu Umakantha Das to Mgt. Tipperah, 10 July 1880, BPP, November 1880, no. 7.

[8] Report on Hill Tipperah for 1876 in Coxhead to Sec. GOB, 26 June 1876, BPP, September 1876, nos 30–1.

phoolbehai,[9] the judges of the High Court found the mother of the son to have been a maid servant, with little direct evidence of any marriage. However, now it was accepted that in the absence of certain male relatives, the 'illegitimate' son by a maid servant, and even of a concubine, may claim the succession to the Raj. However, the price for this admission was the public acknowledgement of 'illegitimacy' and where pronouncements of caste had occurred, to one of 'mixed' caste-ranking. In combination with the onset of the Census operations, such judicial and official pronouncements of 'mixed' jati-rank lent added vigour to ongoing tensions within the group of male kin in the polity concerned. Hence one of the strategies deployed by males whose claims to rule were tenuous, and hard-won over the claims of elder men in the same household, was the greater association with 'Hindu' law and rituals, alongside a deepening and lengthening of genealogies. As in the instance of Tripura, Beerchandra's efforts in this direction established that sovereignty of Tripura was as old as the time of the epic *Mahabharata*; his claim to the title of Maharaj lay in the fact that he was one hundred and seventy-third in descent from that original king.[10] The support such hegemons gave to the Sanskritization/ Kshatriyazation movements within their own states thus began by inviting prominent 'learned pundits of Vikrampore in Dacca' to Agartala, protecting and maintaining them and effectively sponsoring a movement 'for raising their position as Hindoos'.[11] Given the judicial canonization of ritual many ruling males began to invest in elaborate classical 'Sanskritic' marriage-ritual as insurance for the future. Thus, as L.N. Ghose said of the sixteen-year-old ruler of Kuch Bihar, Nripendronarain's marriage to the daughter of Keshub Sen in 1878, 'it was necessary to the legality of the marriage that the rites should be in accordance with the Hindu religion... the fact that Brahmins consented to perform it shows that the marriage was recognised by the Hindus as orthodox'.[12] The prior histories of succession conflicts

[9] Rungadhur Nurendra Mardraj Mahapattur vs. Juggurnath Bhromourbor Roy, in *Shome's Law Reporter*, (Calcutta, 1878), I, November 1877, pp. 94–6.

[10] Beer Chandra Dass Burmon to Sec. GOI, Foreign, 7 February 1874, WBSA, Judcl. (Poll.), July 1874, no. 20.

[11] Asst. Poll. Agent Hill Tipperah to Mgt. Tipperah, 7 July 1882, BPP, November 1882, no. 2.

[12] L.N. Ghose *A Modern History of the Indian Chiefs, Rajas and Zamindars*

in such states and the intervention of the Company's agents determined the direction of social and political change in each instance.

As long as law provided moral legitimation for the continuation of slavery, it was not legislative statute or legal codes that changed the patterns of slave-holding and transfers between different sites in the 'domestic' economy. Moreover, the intellectual and material origins of such regulation were rarely abolitionist. Despite the sustained pressure of the Anti-slavery agitation upon individual Governors-General especially between 1827 and 1853, it is difficult to record the translation of ideological imperatives into administrative fiat. Contrary to the representation of colonial practice in India being directed from the metropolis, the various chapters have shown that colonial initiatives sometimes contradicted metropolitan ones and sometimes preceded and exceeded metropolitan laws. The mode of 'managing' slavery that officials like Prinsep had strongly urged—by executive and piecemeal measures, rather than by law—came to pass. Slavery, and transfers of slaves in regions which had originated the legislation of the period 1811–32, namely Nepal and Assam, were managed entirely by political officers from 1860s to 1870s. Thus in 1867–8, the matter of reversing the trade in children after the famine in eastern India fell to the English Resident at the Darbar in Nepal and various local administrative officers in the Indian provinces. The route for the trade between Nepal and British Indian territories stayed the same as had its composition. Various 'Cutcherries and military posts between the frontier and Khatmandu were... depots for the purchase and sale of human beings' from British India; a 'large proportion' of the 300–400 slaves in Nepal continued to be 'very young children' from regions like Champaran, Balasore and Rangpur.[13] The Commissioner of Assam reported in 1873 that 'slavery was everywhere in fashion';[14] and a large number of Naga and Assamese

(Calcutta, 1881), I, fn., p. 200; for Government assurances to the father of the bride regarding the 'legality' of the marriage, *see* Dy. Commr. Kuch Bihar to Commr., 30 March 1878, BPP, August 1878, no. 19.

[13] Col. G. Ramsay, Resident at Nepal to Sec. GOI, Foreign, 4 April 1867, WBSA, GPP, May 1867, no. 3; Ramsay to Sec. GOI, 21 May 1867, GPP, June 1867, no. 33; J. Beames, Mgt. of Balasore to Sec. GOB, 22 February 1868, GPP, February 1868, no. 38.

[14] AGG North-East Frontier to Sec. GOB, 30 April 1873, WBSA, Judcl. (Poll.) May 1873, no. 38.

slaves continued to be held by the Singhphos in what another official characterized as a 'wheel within wheel' complex of holdings and transfers.[15] To gauge the spread of the phenomenon by the end of the nineteenth century, we need only refer to non-official records of the early twentieth century which suggest that the older patterns of circulation of women and children between various 'domestic' sites continued under the aegis of an expanding rentier aristocracy. In a memorial to the Lieutenant-Governor of Bengal in 1907, the Society for the Protection of Children in India (SPCI) described numerous instances of metaphorical daughterhood and the transfers of such daughters. In one instance, the memorial said, a woman of 'ill fame' died in a hospital, leaving a seven-year old girl. Investigations revealed that the girl was in 'no way related to the woman but had been... procured by her. The child could give no account of herself, having been for some time in the woman's possession'.[16] Another five or six year old girl, found to be living with a prostitute, was also found to be unrelated to her custodian: the woman stated that the child was being reared in order to give her to a man who would then maintain the older woman.[17] Another such 'daughter' aged 2 1/2 years was inherited by two women in succession till a man claiming to be a relative appeared to remove the child and endeavour to sell her in several places in the city. The child was eventually found in the possession of a woman who demanded Rs 25 for the girl alleging that this was what the child had cost her.[18] The transfers occurred not just within the city of Calcutta, but between the city and the outlying areas, as for instance, that of a 3-month-old female 'given' to a prostitute who then placed her with a family in a village, from where she was passed to another village and so on till she died.[19]

The SPCI memorial is important because it helps to complicate the official mythology of 'hereditary' prostitution—how did such labour pass on in the blood when the children were not born of them?

[15] Dy. Commr. Lakhimpore to P.A. to Commr. Assam, 14 April 1873, ibid., no. 39.

[16] Case no. 65, 1903, in J.G.Bell, Sec. of SPCI to President of the Committee to consider the Memorial, 31 October 1907, WBSA, GOB, Home (Poll.), Confidential, File no. 122 of 1907, serial nos 1–4.

[17] Case no. 383, 1907, ibid.

[18] Case no. 131, 1904, ibid.

[19] Case no. 188, 1905, ibid.

Furthermore, the memorial hinted at the stakes that the Babus of Calcutta had in the transfers and transactions around such females. For example, the memorial described one Oriya girl, removed from a brothel to a home, being two years later removed from there by a 'Babu [who] provided the means, and the girl was taken in a *ghari* to a house of ill-fame in the city'.[20] Another member of the SPCI sent to recover another purchased twelve-year-old girl found himself 'mobbed and deprived of the girl, a Babu in Government service taking a prominent part in the proceedings'.[21] Cornelia Sorabji, a lawyer, could encompass the connection of the Babus with the inner slaving economies, for which the middlewomen had borne the brunt. She wrote thus

> Jalpani (food and water or refreshments) is the name given to the arrangement by which rich elderly men import children from villages into a town, and place them in the care of a woman paid for their upkeep—'food and water'—so that they might be ready and secretly accessible to these men whenever required. They are imported at the ages of 2 and 3, and are thrown away as sucked oranges, into the gutters of the common brothels as soon as they mature, or cease to attract their original and individual patrons.[22]

In the face of a constant circulation of female slaves through the households and the markets, the false presumptions on the basis of which divides between marriage, concubinage and prostitution in a polygynous and slave-holding context have been constructed by historians are finally open to re-examination. To quote one of them,

> ... unlike the wife who sells her body 'into slavery once for all', and is made to turn it into a breeding machine for producing sons (...), the prostitute hires out her body without the obligation to reproduce. What she produces in the course of her labour is intangible for herself, but is a purchaseable fantasy for her male client.[23]

Any simple separation of sexual labour (that which produces children or that which produces male fantasies of power) is historically

[20] Case no. 125, 1904, ibid.

[21] Case no, 194, 1905, ibid.

[22] Note on the National Council of Women in India: The Traffic in Women and Children in Sorabji Papers, Mss. Eur. F. 65/155, p. 2.

[23] Sumanta Banerjee, 'The "Beshya" and the "Babu": Prostitute and Her Clientele in 19th Century Bengal', *EPW*, 28, 45, 1993, p. 2461.

untenable. The evidence is that instead of the neat elision of reproductive labour with marriage, and other sexual labours with prostitution, slaves-as-prostitutes, as much as slaves-as-wives, performed the reproductive labours necessary to the production of human wealth (children) and inanimate wealth (houses, cash earnings). Depending on mores that determined status they reproduced lineage, or non-lineage. Constraints of space have prevented a detailed examination of the larger field of sexual labours including those performed by male slaves, but it must be recognized that the representation of all sexual labour as the domain of women is untenable. Finally, in the context of many slave-holders being women, historians need to challenge an *a priori* assumption that the extraction of sexual labour represents the domination of the adult male alone. The system did not allow for gendered tenderness: male and female holders and hirers were equally merciless, or arbitrarily benevolent.

For the reasons outlined above, the rigours of a slave-formation were ameliorated not by law, but by extra-legal institutions like those of the spiritual and religious worlds, and was the result of the activity of the slaves themselves. Much has already been written about slave-resistance—the tiny acts of manipulation, sarcasm, and silence, as well as those portentuous ones closely scrutinized by owners, like flight, poisoning and murder. It was usually a slave in the same household who was instrumental in bringing the murders of slaves to light, a fact well hidden in the *Parliamentary Papers*; but recounted in other contemporaneous sources. Hence when in July 1837 a severely wounded eight-year-old slave girl was sent off from Calcutta to Chinsurah in the charge of an African slave-eunuch of the household, the latter diverted the palki to the police-office. The murder of a thirty-one year-old slave woman was denounced by another male slave belonging to the same household, in revenge upon his mistress who had taken away a child of his 'as her property'. However, resistance too was no simple matter. For every slave who helped another slave to run away, there were slaves policing other slaves on behalf of the master or mistress as in the case of the eunuchs in the Nizamat, sometimes killing them at the behest of the latter, or acquiring other slaves for themselves or their masters and mistresses. For every slave who refused the visage of compliance, there were others who could not do so overtly.

Because slavery had been imagined as a problem in coercion and labour and not as a problematic dialectic between alienation and 'belongingness', the significance of religious and other hegemonic expressions of belonging (identity) for slave-resistance and slave-mobility is little understood. Much historical ink has been spilt on denying the connection between devout Islam and slavery; this denial is based upon the attribution of egalitarian desires to slaves, and the continuation of differentiation within the Islamic congregation. Yet, evidence of the desire for 'egalitarianism' in the breast of slaves is yet to be established. On the other hand, the physical presence of slaves within the Islamic fold is undeniable. Explaining the latter requires imagining the *jama'at* (or assembly) as an experience in belonging *with* a wider group, and not *to* a household alone. Belonging with a devotional group may have compensated for the absence of other corporate identities and networks. In fact, especially after the 1870s when a classicization of Islamic practice occurred in Bengal, the promise held out by the Quran, was not egalitarianism, but freedom to slaves in return for devoted service to earthly and divine masters.

Furthermore, religious beliefs and the moral economy had had a subtle and long relationship with issues of political economy within which slaves found themselves. In a context where proprietory title to immoveables like land and buildings was denied to the slave and ex-slave by erstwhile owners, the gifting of little parcels of income/lands/slaves to a God/Supreme Being could inflict temporal masters with partial deprivation, and reserve for slaves a tiny portion of the gains of their labour in the form of the deference of the populace that prayed at these shrines.[24] It was one form of being remembered, in a social formation which paid homage to ancestors. By the act of donating from their own earnings to a higher God, successful slaves salvaged some pride and dealt a blow to the masterly conceit that slaves only received goods, maintenance, money.[25] The establishment of mosques and *madrassahs* on specific grants of lakhiraj land for religious and charitable purposes was a strategy many a master would have found

[24] Memorial of Mir Jaffer Ali of Jaffarganj, Abdul Qadir of Dowlat Madar, Feda Ali of Ichagunj, Mirza Qasim Ali and Mirza Enayet Ali of the city of Murshidabad, 6 June 1896, WBSA, Judcl., September 1896, no. 28.

[25] *See* Carl F. Petry, 'From Slaves to Benefactors: The Habashis of Mamluk Cairo', *Sudanic Africa*, 5, 1994, pp. 57–66.

hard to contest. The eunuch Basant Ali Khan was only one of the slaves of the Nizamat whom we have studied in detail: other slave-eunuchs like Amān Ali Khan and Dārāb Ali Khan also left substantial estates for devotional purposes. Such donations for spiritual benefit were important to slaves. As Marmon had explained for Mamluk formations, the association of particular eunuchs with the Prophet's tomb at Mecca represented an elevation whereby senior eunuch 'servants of the earthly prince' were often freedmen, servants of a more exalted order. By the same token, lakhiraj tenure-holders in slave-holding households like Murshidabad, Dacca and elsewhere often had the prior jural status of slave-servant of an earthly master, who had moved to the service of a higher god: something not yet thoroughly examined by historians of endowments in colonial India.[26] Furthermore, the evidence from the Nizamat suggests that not only were the slave-chelas trained by particular eunuchs reassigned as servants to shrines, but that the major issue for such servants then became the assumption of authority as freedmen proprietors of those lands which yielded revenues for the maintenance of the shrine itself. On the death of the father/original owner of Babbu Begam, the latter assigned one male slave as *mujawir* (literally, fixed to the temple) and five as qaris to the tomb of Muhammad Sami Khan. For the maintenance of the servants of the tomb, she purchased five *beegahs* of land which formed a part of the Nizamat Ganjiyat. By 1844, however, the heir Amir Ali Khan, reported that the mujawir, apart from a 'want of respect' to the heir of the family, had begun to conceive 'that he is the real owner of the land and as such perfectly independent. Instead of reserving a part of the proceeds for repairing the tombs, he spends the whole as if... he is answerable to no one for his acts' while continuing to receive a monthly sum of Rs 28 from the de'orhi of Amīrunnisa Begam.[27]

If this was true of slave men and eunuchs, it was also true of female slaves. Pilgrimage, as well as service at shrines and tombs associated with erstwhile masters, were important to concubines and other slaves. However, since the issue of female service at temples attracted special

[26] *See* Gregory C. Kozlowski, *Muslim Endowments and Society in British India* (Cambridge, 1985).

[27] Amir Ali Khan to AGG Maj. Gen. F.V. Raper, 12 July 1844, *MNLR*, II, p. 495.

attention and vilification in the twentieth century, attempts to investigate the interlinkage between manumission and service to a higher god run the risk of crude simplifications. Nevertheless, where female slaves were themselves 'gifted' to a god, there too the metaphor remained that of affinity and marriage. Female slaves were described as having been 'married' to the deity. This phenomenon, well studied now by Marglin and others,[28] appears to have been common to the Muslim syncretic tradition in inner Bengal as well. At least all the *sukris* who were rounded up in the course of an investigation into 'mock marriages' in 1872, proclaimed that they had been married to 'Gazee Meah', a bamboo pole worshipped as a deity in Bankura: this marriage, they believed, removed them from mortal machinations. 'It would be a great sin for us to marry a man', as one of them put it.[29] Celibacy, and alms-collecting, that such ceremonies committed the seven-and-eleven-year-olds to, appeared to have called forth special enactments to stop such marriages. Yet like so many other colonial gestures of this kind, such proclamations missed the more fundamental changes occurring in the metaphysical and religious worlds of many men and women in Bengal from the seventies. That was part of the story of classicization of Islam in Bengal, a process wherein the moral imperatives of Quranic prescription about the manumission of slaves, the proscription of putting female slaves to work as prostitutes for the profit of their holders, and advice to poor men to marry such freed female slaves was propagated by itinerant and poor maulvis and translated into the everyday actions of small knots of rural Musalmans. By the end of the nineteenth century, the movement to 'purify' Islam of deviations from Quranic law impinged centrally on the definition of nikah. All these re-definitions were critical particularly for female slaves. For instance, in 1906, Maulvi Samīruddīn had preached that 'the practice of Hindu zamindars of keeping prostitutes in the bazars was a bad one, and someone had retorted that the prostitutes were all Muhammadans. He had replied that disgrace should be removed from

[28] F.A.Marglin, *Wives of the God-King: the Rituals of the Devadasis of Puri* (Delhi, 1985); Janaki Nair, 'The Devadasis, Dharma and the State', *Economic and Political Weekly*, 29, 50, 1994, pp. 3157–67.

[29] Cited in Report by W.L.F. Robinson, Commsr. of Rajshahi, 2 August 1872, WBSA, Judcl., October 1872, nos 95–7.

the Muhammadan community, and the women should be taken away and married'.[30] A spate of manumissions, without payment of the customary ransom, appear to have occurred, which colonial officials reported as the actions of 'low-class Muhammadans' removing and forcibly releasing Muhammadan 'servants' from their Hindu masters, including 'prostitutes' from the *hat* of the taluqdars of many sub-divisions of Bengal, like Mymensingh, Bakarganj, Rangpur. Or as in Jamalpur in April 1907, buildings that encapsulated all aspects of a political economy based on female slave-use and hire, the *nachghars* (halls of dance) itself, were demolished at fairs (*melas*) and bazars belonging to various under-tenants and zamindars. However, since many of the latter had also begun to dominate the institutions of the police and the judiciary, these forcible manumissions increasingly became part and parcel of the growth of communal and nationalist politics in Eastern India through the stereotype of the 'rapacious' Muslim male *budmash*.[31] Such contests were heir to a long tradition of slave-enticement cases of the eighteenth and nineteenth centuries. Though framed as contests between volition (of the female) and coercion (by males), they were principally about belonging, about identity, and about status.

At this moment in the history of the region, when one hears much talk of a uniform civil code, it is necessary to remind ourselves that what is being battled with are personal laws born after peculiar colonial gestations. To reserve venom for fundamentalist Islam is to ignore a wider historical dilemma faced by many religions, satisfying the yearnings of slaves and ex-slaves, as well as providing legitimation to their masters. To understand that historical condition within which revivalist and fundamentalist faiths grew in nineteenth century India, historians are called upon to unwork a history of slavery from within the palimpsest of genealogy, kinship and colonial law, and to rework

[30] Magt. of Mymensingh to Commr. Dacca Divn., 1 June 1906, in L/P&J/6/826, file 3081, enclosure 1.

[31] Pradip K. Datta, '"Dying Hindus": the Production of Hindu Communal Common Sense in Early 20th Century Bengal', *Economic and Political Weekly*, 28, 25, 1993, pp. 1305–19; Ritu Menon, 'Recovery, Rupture, Resistance: Indian State and Abduction of Women During Partition', *Economic and Political Weekly*, 28, 27, 1993, pp. WS2–WS11 and Urvashi Butalia, 'Community, State and Gender: On Women's Agency During Partition', ibid., pp. WS12–WS24.

the hard labour of sexual reproduction into assessments of historical labour systems. What is needed is not so much a history from below, but a history from within, a history of exile from natalities and the recovery of belonging, a history in which nothing was too ephemereal and no grain of sand—the annas and pice of a slave-girl's peculium, or the petty fines paid by a slave-concubine—too tiny to make up the bedrock of a colonial empire. Only then can the legacy of administrative complicity and familial collusion in the persistence of a traffic in women and children in modern India be fully understood.

Appendix I

I. Table of Claimants in the Property of Shahamat Jang

List No. I, (the sons)	List No. II, (Ruheemunessa)
Izzatulnessa, mother of Mīr J'afer Ally (about 25 years old), Syed Ally (about 20 years old) and of Ameena Begam (about 16 years old).	Toofah, a Moglanee purchased by Fukeerutnessa, wife of Kauzim Ally Khan; Fukeerut's daughter was S.J.'s first wife and her mother gave her Toofah on the occasion of her marriage.
Hoormutulnessa Begam, decd., mother of Bubber Ally (about 18 years old), Hadjee Begam (about 26 years old) and Momenah Begam (about 24 years old).	Muneejah, stands on the same footing as Toofah.
Sherifulnessa alias Shurfun, mother of Turbeatulnessa Begum (about 4 years old).	Shurfun, a slave purchased by S.J.
Sherulnessa alias Hingun, mother of Rahutulnessa Begam (about 5 years old).	Hingun was brought up and protected by S.J.'s first wife.
Shumsulnessa alias Hirmeen, no issue by her.	—
Zeenutnessa alias Shubraty, mother of Meer Mohummud Ally (about 5 years old).	Shubratee, a present from Bubboo Begam to Afzulnessa Begam, second wife of S.J.

List No. I, (the sons)		List No. II, (Ruheemunessa)	
Kumurulnessa alias Begumjee, no issue by her.		Begumjee alias Begumee, brought from Maldah to reside in S.J.'s house as a Concubine.	
Khadimah:	Immuneelah	Purchased by S.J.:	Immuneelah
	Premcullee		Penculla
	Fuckeerun		Fuckeerun
	Bhālee		Bhalee
	Munnah		Munneah
	Mooneah		Mooneah
	Rostum		Roshun
	Feekah		Pheekah
	Golaubee		Golaubee
	Mauncoohar.		Mauncooh

Source: Letter of Superintendent of Nizamat Affairs, 2 September 1809, in OIOC, B.C., F/4/312/7131. Spellings of original source.

II. Table of Claimants in Property of Rustam Jang

A. Sharokh Begam, *munkooha,* and Noorjehan Begam, daughter.
Soortellee, aka Boo Begami Kulaun, *mumtooah by nikah;* her son Meer Mohummad Ali, 5 years old; her daughter Qutbunnissa Begam, 6 years old; her daughter Umutul Bootool Begam, 7 years old.
Daoodee, aka Boo Begami Khurd, *mumtooah by nikah.*
Muhbooba, Abyssinian, *nikah o'mutaa.*
Shereen, *nikah o'mutaa.*
Ufrozee, *nikah o'mutaa.*

B. Dependents and domestics: Gooldista, Goolzar, Chumpa, Lallkuar, Utcheea, Zun-i-nigar, Chumpa 2, Bahar Khanum, Behadur Chela, Jeetun Chela.

Source: Letters of Superintendent of Nizamat Affairs, 10 July 1809 in BPC, P/118/43 and 26 March 1810 in OIOC, B.C., F/4/333/7635. Spellings of original sources.

Table II. Deori of Badrunnisa Begam, circa 1850.

Names & service	Length of service	Age in 1857	Age at beginning	Monthly pension recommended
Fyzun urf Ameerun, mother of Baker Ali	—	40 years		25–0–0
Hosainee, Moghulanee	25 years	40 years	15 yr–old	15–0–0
Bunnee, Moghulanee	19 years	37 years	18 yr–old	5–0–0
Boo Shurfoo, khawas	50 years	56 years	6 yr–old	5–0–0
Booboo Phirkah, khawas	35 years	45 years	10 yr–old	5–0–0
Booboo Sudda Kunwar, khawas	58 years	70 years	12 yr–old	4–0–0
Gaffun, khawas	28 years	41 years	13 yr–old	4–0–0
Kurumbuksh, khawas	33 years	44 years	11 yr–old	4–0–0
Ruttun, khawas	51 years	66 years	15 yr–old	4–0–0
Fyzun, khawas	29 years	42 years	13 yr–old	4–0–0
Dhoomun, khawas	28 years	39 years	11 yr–old	4–0–0
Hoorun, khawas	26 years	40 years	14 yr–old	4–0–0
Bhorun, khawas	30 years	44 years	14 yr–old	3–0–0
Jaddoh, khawas	31 years	45 years	14 yr–old	3–0–0

Names & service	Length of service	Age in 1857	Age at beginning	Monthly pension recommended
Bunneejan, khawas	20 years	32 years	12 yr–old	3–0–0
Doordanah, khawas	29 years	41 years	11 yr–old	3–0–0
Burrattee, khawas	35 years	44 years	9 yr–old	3–0–0
Beebun, khawas	23 years	74 years	54 yr–old	4–0–0
Sahebjan, luknee	14 years	40 years	30 yr–old	2–8–0
Zeebun, dhobin	13 years	60 years	47 yr–old	3–0–0
Horoh metranee	18 years	45 years	27 yr–old	4–0–0
Haji Abdul Rasool, Karee al Imambara	20 years	44 years	24 yr–old	5–0–0
Mirza Kader Ali, Karee al Masjid	10 years	22 years	12 yr–old	5–0–0
Syud Haji Kurban Ali, Karee	14 years	46 years	34 yr–old	5–0–0
Shaikh Bukaollah, karee	23 years	41 years	19 yr–old	2–0–0
Sujatollah, dandee	14 years	60 years	46 yr–old	3–0–0
Hosainbuksh abdar	32 years	45 years	13 yr–old	3–0–0
Ramloll, dheulat?	32 years	45 years	13 yr–old	4–0–0

Source: BC F/4/2708/194266. Spellings of original sources.

Appendix II

Document I: (English text of the deed shown on cover)

In consideration of the sum of Eight Arcot Rupees/a Rs 8 to me in hand paid, I Jumat Cawn Haoldar of Calcutta do hereby make sale Assign and Deliver and do by these presents forever quit Claim of my Slave Boy named Dowlat about Nine year unto Mr. Lochemey, In witness whereof I have here unto Set my hand, In Calcutta Collector's office this 10 day of November 1764.

Registered by (Signature illegible)

Collr.

The Bengali text refers to this document as *bikrayapatram* (document of sale), and to the subject of the transaction as *amar kharida ek chhokra naam dowlut boyesh noye bochhor,* specifying age, name, and the mode in which the seller had acquired the slave. The Bengali text also refers to the volition of the seller, as well as a guarantee on behalf of the seller's heirs and descendants: *ami apon khosh bechai tomar sthane nogod dokkhino arkat at taka paiya bikroye korilam... ei chhokrar shohit amar kono dawa nai... ami o amar waris keho dawares ebong dawadar... ei bayane...* (I have voluntarily sold to you for the ready money of eight Arcot Rupees... relinquishing all claims upon this slave/chhokra... on behalf of my self and my heirs and successors hereafter)

Table I: Decisions on enticement.

Number and date	Prosecutor	Prisoner and charge	Sentence by Daroga and Maulavis of the Faujdari Adalut
14 March, 1774 no. 363.	John Baptiste	Durbaree & Betty for enticing away and selling the plaintiff's slave-boy.	Durbaree sold the boy to Betty. Durbaree to receive 5 Corah and return the boy.
12 April, 1774 no. 408	Sunker	Goupie Ramkishore and Isserry, for enticing away the ptff.'s slave-girl.	Issery found guilty, ordered to received 10 'shoes', Goupee and Ramkishore not guilty.
20 April, 1774 no. 422.	Johanna	Bockull, for enticing the pltff's slave-girl.	Not guilty. The Girl went away of herself. Ordered she receive 10 'latts'.
9 May, 1774 no. 460.	Ramcukar	Ariff, for enticing Golaf the slave-girl.	Proved that Ariff enticed away Golaf, ordered to be punished with 30 latts round the Town & repay what Jewels were in Golaf's possession. Golaf to receive 10 latts.
28 May, 1774 no. 501.	Thomas	Sirdar Bearer, for enticing away pltff.'s slave-girl.	Proved. To receive 15 Rattans.
12 June, 1774 no. 530.	Mahmud Warris	Chanulla & Sunulla, for enticing way the pltff's slave-girl.	Chunulla to be confined 10 days, Sunulla to receive 10 latts.
29 June, 1774 no. 556.	Totteram & Salemon	Mohon Singh, for running away with the pltff's slave-girl.	To be punished with 5 'Chabucks' and to return the girl in 2 days.
20 July, 1774 no. 603.	Taramunny	Neerusam, for taking the pltff's slave-girl.	The girl belongs to the defendant. Ordered that he give the plaintiff 10 Rupees & the jewels the girl had.

Number and date	Prosecutor	Prisoner and charge	Sentence by Daroga and Maulavis of the Faujdari Adalut
24 July, 1774 no. 612.	Sybcawn	Callee, for enticing away pltff's slave-girl.	Guilty-to be punished with 15 Rattans.
23 Aug., 1774 no. 688.	Mr. Petrie	Joanna & Emamdie, for enticing away pltff's slave-girl	Emamdie to be punished with 15 Rattans round the town, Joana to be confined in Bridewell during the pltff's pleasure.
10 Sept., 1774 no. 720	Treasa	Suda Cawn, for enticing away pltff's slave-girl	To be punished with 5 Rattans.
29 Oct., 1774 no. 802.	Chandah	Ramnarain Doss, for taking the pltff's slave-girls.	Proved the Girls are the Defendant's Property. Ordered he have them.
31 Oct., 1774 no. 806.	Totteram	Cunchee, for taking away the pltff's slave-girls.	It appears the slave girls belong to the pltff's wife. Ordered the Defendant to return the slave girls to the pltff. as his wife's property.
23 Nov., 1774 no. 822.	Cornee	Tilleram, for enticing pltff's slave-girl.	Ordered prisoner 5 'shoes'.
6 Dec., 1774 no. 835.	Daniel Empson	Tom, for enticing away pltff's slave-girl.	Not guilty. It appears the Girl was running away of her own accord. Ordered her 3 lashes.
28 Dec., 1774 no. 858.	Goverdon Tagoor.	Ramdoss & Biragees, for enticing away pltff's slave-girl with jewels and goods.	It appears that 5 Bairagues were kept in the pltff's house in his absence by pltff's mother & the defdt. was security for them, that they ran away with the pltff's Girl & goods. Ordered the defdt. to produce the offenders.

Number and date	Prosecutor	Prisoner and charge	Sentence by Daroga and Maulavis of the Faujdari Adalut
29 Dec., 1774 no. 862.	Annoo	Tibboo & Suruboy, for enticing away pltff's slave-girl.	Not guilty–appears that that Tibboo is the pltff's slave, and that Saruboi is her godfather therefore acquit him; ordered the pltff. to produce the slavery paper in which have no mention of her name or those of her Father and mother-She is ordered to return to her godfather and remain a free woman.

Source: Proceedings of the Faujdari Adalut for 1774, P/154/39.

Table II: Decisions of entitlement.

Date and Number	Prosecutor	Defendant and charge	Sentence by the Faujdari Court
Jan. 1774, no. 250	Jumuna	Mogalishram, for taking her son.	Proved that the pltff. sold the child to the defdt.—ordered that the defdt. shall have the boy.
4 March, no. 344	Sookmoy	Anundee, for endeavouring to make the pltff. a slave	Proved that the pltff. is defdt.'s slave
31 March, no. 398	Conchunni	Moias, for making the pltff. a slave	Proved that pltff. is defdt.'s slave—ordered to go to his house.
31 March, no. 400	Tattoo	Zettoo, for endeavouring to make the pltff. a slave.	The pltff. is not the defdt.'s slave to go where he pleases.
3 Aug., no. 636	Tom	Duneego, for making the pltff. a slave	The pltff. is the defdt.'s slave, ordered to go to his master.

Date and Number	Prosecutor	Defendant and charge	Sentence by the Faujdari Court
6 Aug., no. 649	Cammoo	Peggy, for taking the pltff.'s wife.	The woman is the defdt.'s slave & not the pltff.'s wife. Ordered that the defdt. have her.
7 Sept., no. 713	Caundoo	Sorphon, for not staying in the pltff.'s house.	The defdt. was the pltff.'s slave but agreed with her master to pay him Rs. 200 for her liberty to which Bargain he also agreed. Ordered to pay Rs. 200 and she to have her liberty.
26 Oct., no. 796	Sonamuckee	Rotton slave, for running away.	The defdt. is not the pltff.'s slave. Ordered the pltff. to pay Rs. 26 to defdt. and go where she pleases.

Source: Compiled from Progs. of the Faujdari Adalut, 1774, P/154/39. Spellings of original source.

Bibliography

Private Papers and Manuscripts

British Library, London.

Auckland Papers.

Broughton Papers.

Hastings Papers.

Wellesley Papers.

Oriental and India Office Collection, London

Adam Papers.

Bentinck Papers.

Marathi, D. 31

Orme Collection.

Sorabji Papers

Verlee Collection.

National Library, Calcutta

JN Sarkar Collection, 174.

Published Official Records

An Abstract of the Regulations Enacted for the Administration of the Police and Criminal Justice in the Provinces of Bengal, Bihar and Orissa (ed. W. Blunt and H. Shakespear, Calcutta, 1824).

A Collection of Statutes Relating to India (2 vols, Government of India, Calcutta, 1881).

Calendar of Persian Correspondence, (11 vols, Calcutta, 1911 Imperial Records Deptt., New Delhi, National Archives of India 1953).

Circular Orders Passed by the Nizamut Adalut and Communicated to the Courts of Circuit & Zillah & City Magistrates by the Register of that Court (2 vols, Calcutta, Bengal Military Orphan Press 1829–38).

Circular Orders Passed by the Nizamut Communicated to the Criminal Authorities from 1796 to 1853 (ed. J. Carrau, Calcutta, Bengal Military Orphan Press, 1855).

Constructions by the Courts of the Sudder Dewanny and the Nizamut Adawlut of the Regulations and Laws for the Civil Government of the Whole of the Territories under the Presidency of Fort William in Bengal (2 vols, Calcutta, Bengal Military Orphan Press, 1839–43).

Fort William—India House Correspondence.

Indian Records with a Commercial View of the Relations Between the British Government & the Nawabs Nazim of Bengal Behar and Orissa (first ed. 1870, reprint, Delhi, New Age Publications 1985.)

List of the Private Secretaries to the Governors-General and Viceroys from 1774 to 1908 (Calcutta, Government Superintendent of Printing, 1908).

Moore, Edmund F. *Reports of Cases Heard and Determined by the Privy Council on Appeal from the East Indies* (14 vols, Reprint, Calcutta, 1915).

Papers relating to East India Affairs: Regulations passed between 1810 and 1824 (London, 1825).

Press List of Ancient Records of the Government of India in the Public Department.

Records of the Government of Bengal: The Letter Copy Books of the Resident at the Durbar at Murshidabad, 1769–1776 (ed. Walter K. Firminger, 3 vols, Calcutta, 1919).

Regulations Passed by the Governor-General in Council (3 vols, London, 1828–29).

Report on Cases Decided in the Nizamut Adalut (Calcutta, 1827–56).

Report on Cases Decided in the Sudder Dewanny Adawlut (Calcutta, 1827–65).

Selections from Unpublished Records of Government for the years 1748–1767 (ed. J. Long, Calcutta, 1869).

Selection from the Indian Despatches, Memoranda &ca. of the Duke of Wellington (ed. S. J. Owen, Clarendon Press, 1880).

Selections from the State Papers of the Governors-General of India (ed. G. Forrest, Oxford, Basil Blackwell, reprint 1980).

Selections from the Records of the Bombay Government: Sketches of the Native States (Bombay, 1856).

Selections from the Satara Raja's and Peshwa's Diaries (9 vols Pune, Deccan Vernacular Translation Society 1902–11).

Selections from the Peshwa Daftar (ed. G. S. Sardesai, Bombay, Govt. Central Press 1931–34).

Sutherland's Weekly Reporter (Calcutta, 1861–80).

Sylhet District Records (ed. W.K. Firminger, 3 vols, Shillong 1913–16).

The Indian Penal Code as Originally Framed in 1837 (Reprint, Madras 1888).

Twenty Five Questions Addressed to the Rajahs & Chiefs of the Regulation and Tributary Mehals by the Superintendent in 1814 (third ed. Calcutta Bengal Secretariat, 1905).

West Bengal District Records: New Series Burdwan: Letters Received, 1788–1802 (ed. A. Mitra, Calcutta, 1955).

Murshidabad: Letters Received, 1789–1803. (ed. S. K. Bose and A. Mitra, Calcutta, 1958).

Murshidabad Nizamut: Letters Received, pt. I, 1793–1856. (ed. J. Datta Gupta and S. K. Bose, Calcutta, 1964).

Murshidabad Nizamut: Letters Issued, pt. I, 1802–1831. (ed. J. Datta Gupta and S. K. Bose, Calcutta, 1965).

Murshidabad Nizamut: Letters Issued, pt. II. 1834–1872. (ed. J. Datta Gupta and S. K. Bose, Calcutta, 1967).

Murshidabad Nizamut: Letters Received, pt. II. 1807–1855. (ed. J. Datta Gupta and S. K. Bose, Calcutta, 1969).

Parliamentary Debates (Hansard, London, 1833), XIX– XXI.

Parliamentary Papers:

1828 (House of Commons), vol. 4, no. 125: Slavery in India: Correspondence and abstract of Regulations from 1772.

1834 (House of Commons), vol. 44, no. 128: Slavery in India: Correspondence.

1837–38 (House of Commons), vol. 51, no. 697: Slave Trade: Correspondence, Orders, Regulations and Proceedings thereon, 1829–36.

1839 (House of Commons), no. 138.

1841 (House of Commons), vol. 28, no. 262: Slavery in India: Report of the Law Commissioners with Appendix.

1856 (House of Commons), vol. 65. Minute of the Marquis of Dalhousie, 28 February 1856, on the Administration of India.

1860 (House of Lords) Session II, no. 272: Judicial Despatch to India on the Leasing Out of Girls.

1871 (House of Commons), vol. 99. East India: Home Accounts of the Government of India; Financial Correspondence on the Deposit Fund of the Nawab Nazims of Murshidabad.

Unpublished Official Sources, OIOC.

Bengal Board of Revenue Consultations, 1774–1820.

Bengal Board of Revenue, Miscelleneous (Wards) Consultations, 1770–96.

Bengal Criminal Judicial Consultations, (Lower Provinces), 1793–1834.

Bengal Financial Proceedings, 1860–80.

Bengal Judicial Proceedings, 1859–88.

Bengal Political Consultations and Proceedings, 1793–1888.

Bengal Proceedings of the Faujdari Adalut, 1774.

Bengal Procedings of Sadr Diwani Adalut, 1774–1800.

Bengal Public Consultations, 1757–1846.

Bengal Secret & Military Consultations, 1770–90.

Bengal Wills and Inventories, 1770–1880.

Board's Collection, 1793–1855.

Correspondence between Government of Bengal and the Court of Directors, 1800–30.

Home (Miscellaneous) Proceedings.

India Foreign and Political Proceedings, 1834.

India Judicial Proceedings, 1835–75.

India Legislative Proceedings, 1835–46.

India Political Procedings, 1835–75.

Unpublished Sources, National Archives

Foreign General

Foreign Political Proceedings, 1839–99.

Legislative Proceedings, 1840–41.

Unpublished Sources, West Bengal State Archives

Bengal General (Political) Proceedings, 1858–99.

Bengal Judicial Proceedings, 1870–90.

Home (Political) Confidential Proceedings, 1905–09.

Newspapers and Periodicals

Bengal Hurkaru, 1830–44.

Daily News, 1838–43.

Englishman, 1880–3.

Itihas Sangraha

Moorshedabad News, 1838–9.

Journal of the Asiatic Society of Bengal, 1834–78.

Secondary Sources

Adam, William *The Law and Practice of Slavery in British India in a series of Letters to Thomas Fowell Buxton* (London, 1841).

——— *Reports on the State of Education in Bengal* (1835–1838) Anathanath Basu (ed.) (Calcutta, the University, 1941).

Ahmad, Qeyamuddin 'A Mid-Nineteenth Century Case of a Long-Term Lease, Not Sale of Human Beings', *Indian Historical Review*, 15, 1988–9, pp. 276–80.

Ahmed, Amanatullah 'Slavery in North East India', *Indian Historical Records Commission*, 23, 2, 1946.

———*History of Cooch Behar* (in Bengali) (Calcutta, Modern Book Agency, 1990)

Ahmed, Imtiaz (ed.) *Family, Kinship and Marriage Among Muslims in India* (New Delhi, Manohar, 1976).

Ahmed, Leila *Women and Gender in Islam: Historical Roots of a Modern Debate* (New Haven, Yale University Press, 1992).

Ahmed, Rafiuddin *The Bengal Muslims 1871–1906: A Quest for Identity* (Delhi, Oxford University Press, 1988).

Ahuja, Ravi 'Labour Relations in an Early Colonial Context: Madras, 1750–1800', Paper for Conference on Indian Labour History, Delhi, 1998.

Alam, Muzaffar 'Trade, State Policy and Regional Change: Aspects of Mughal Uzbek Commercial Relations, c. 1550–1750', *Journal of the Economic and Social History of the Orient,* 37, 3, 1994, pp. 202–27.

Al-Azmeh, Aziz (ed.) *Islamic Law: Social and Historical Contexts* (London, Routledge, 1988).

Ali, M. Athar *The Mughal Nobility under Aurungzeb* (Bombay 1966, revised edn, Delhi, Oxford University Press, 1997).

———'Elements of Social Justice in Medieval Thought' Indian History Congress, 57th session, 1996 (Delhi, 1997).

Ali, Mrs. Meer Hassan *Observations on the Mussalmans of India* revd. and ed. W. Crooke (reprint Delhi, Deep Publications 1975).

Ali, Shanti Sadiq *The African Dispersal in the Deccan from Medieval to Modern Times* (Hyderabad, Orient Longman, 1996).

Anderson, Mary M. *Hidden Power: the Palace Eunuchs of Imperial China* (Buffalo, Prometheus Books, 1990).

Anderson, Michael R. 'Islamic Law and the Colonial Encounter in British India' in Chibli Mallat and J. Connors (eds) *Islamic Family Law* (London, Graham and Trotman, 1990), pp. 205–23.

———'Work Construed: Ideological Origins of Labour Law in British India to 1918' in P. Robb (ed.) *Dalit Movements and the Meanings of Labour in India* (Delhi, Oxford University Press, 1993), pp. 87–120.

[Anon.] *A Vindication of Mr. Holwell's Character from the Aspersions Thrown Out in an Anonymous Pamphlet of 1764* (London, 1764).

[Anon.] *Slavery and the Slave Trade in British India; with Notices of the Existence of these Evils in the Islands of Ceylon, Malacca and Penang* (London, Thomas Ward and Co., 1841).

Arasaratnam, Sinhappa *Merchants, Companies and Commerce on the Coromandel Coast, 1650–1740* (Delhi, Oxford University Press, 1986).

Archer, Leonie J. (ed.) *Slavery and Other Forms of Unfree Labour* (New York, Routledge, 1988).

Ayalon, David *Studies on the Mameluks of Egypt 1250–1517* (London, Variorum Reprints, 1977).

———*Islam and the Abode of War: Military Slaves and Islamic Adversaries* (London, Variorum, 1994).

Baer, Gabriel *Studies in the Social History of Modern Egypt* (University of Chicago Press, 1969).

Baillie, Neil B.E. *The Mohummudan Law of Inheritance according to Aboo Huneefa and his followers* (London, 1874, 2nd. ed.).

———*Digest of Moohummudan Law, Containing the Doctrines of the Imameea Code of Jurisprudence on the Most Important of the Same Subjects*, (London, Smith, Elder and Co. 1869, second edition 1869).

———*Digest of Moohummudan Law on the Subjects to which it is applied by British Counts of Justice in India* (London, Smith, Elder and Co., 1865).

Balzani, Maria and Varsha Joshi 'The Death of a Concubine's Daughter', *South Asia Research*, 14, 2, 1994, pp. 136–62

Banaji, D. *Slavery in British India* (Bombay, 1933).

Bandopadhyaya, Brajendranath *Sambad Patre Sekaler Katha* (2 vols, Calcutta, 1933)

———'The Mother of the Company', *Bengal Past and Present*, 32, 63–64, 1926, pp. 38–48.

———*Banglar Begum* (Calcutta, 1913).

Banerjee, D.N. 'The Accession of Nazm-ud-Dowla to the Throne of Bengal and the Position of the East India Company', *Bengal Past and Present*, 60, 1941, pp. 19–34.

Banerjee, N.N. *Monograph on Cotton Fabrics of Bengal* (Calcutta, 1898).

Banerjee, Sumanta 'The "Beshya" and "Babu" : Prostitute and Her Clientele' *Economic and Political Weekly*, 28, 45, 1993, pp. 2461–75.

Barnett, Richard B. *North India between Empires: Awadh, the Mughals and the British 1720–1801* (Delhi, Manohar, 1987).

Barooah, N.K. *David Scott in North-East India 1802–1831* (New Delhi, Munshiram Manoharlal, 1970).

Baruah, S.L. 'Sadar Aminar Atmajivani', *Bengal Past and Present*, 97, 1, 1978, pp. 79–85.

Basu, Basanta Kumar 'Notes on Slave Trade and Slavery in India during the Early Days of John Company', *Muslim Review*, 4, 4, 1930, pp. 21–34.

Baxi, Upendra 'People's Law in India, The Hindu Society' in Masaji Chiba (ed.), *Asian Indigenous Law: In Interaction with Received Law* (London, KPI, 1986), pp. 216–66.

Bay, Edna G. 'Belief, Legitimacy and the *Kpojito*', *Journal of African History*, 36, 1995, pp. 1–27.

Bayly, C.A. *Rulers, Townsmen and Bazaars: North Indian Society in the Age of British Expansion 1770–1870* (Cambridge, Cambridge University Press, 1983).

Bayly, Susan '"Caste" and "Race" in the Colonial Ethnography of India' in P. Robb (ed.), *The Concept of Race in South Asia* (Delhi, Oxford University Press, 1995), pp. 165–218.

Beaufort, F.L. *A Digest of the Criminal Law of the Presidency of Fort William and Guide to all the Criminal Authorities Therein* (2 vols Calcutta, 1857).

Beachey, R.W. *The Slave Trade of Eastern Africa* (London, Rex Collings, 1976).

Beckles, Hilary McD, *Natural Rebels, A Social History of Enslaved Black Women in Barbados* (London, Zed Books, 1989).

Bernier, Francois *Travels in the Mogul Empire* A. Constable (trans.), revised by V. A. Smith (Milford, Oxford University Press, 1916).

Bevan, H. *Thirty Years in India or a Soldier's Reminiscences of Native and European Life in the Presidencies from 1808 to 1838* (London, 1839).

Beveridge, H. 'The Mother of Jahangir', *Journal of the Asiatic Society of Bengal*, 56, 1, 1887, pp. 164–7 and 57, 1, 1888, pp. 75–6.

Bhalla, P.N. 'The Mother of the Company', *Journal of Indian History*, 22, 2–3, 1943, pp. 128–44.

Bharatchandra, Ray Gunakar, *Granthavali*, Brajendrachandra Bhattacharya (ed.) (Calcutta, Modern Book Agency, 1961).

Bhattacharya, Ananda 'Sanyasi and Fakir Uprising in Bengal in the Second Half of the Eighteenth Century', unpublished Ph.D. Diss., University of Jadavpur, Calcutta, 1992.

Bhattacharya, Ashutosh *Gopichandrer Gan* (Calcutta, Calcutta Book House, 1959).

Bhattacharya, Jogendranath *An Exposition of the Origins of the Hindu Caste System and the Bearing of the Sects Towards Each Other and Towards Other Religious Systems* (first ed. 1896, reprint Calcutta, 1973).

Biker, Julio F. J. *Colleccao de Tratados e Concertos de Pazes* (10 Vols, Lisbon, 1881–85).

Bissoondoyal, U. and S.B.C. Servansingh *Slavery in South West Indian Ocean* (Mauritius, Mahatma Gandhi Institute, 1989).

Blackburn, Robin 'Defining Slavery—Its Special Features and Social Role' in Leonie J. Archer (ed.), *Slavery and Other Forms of Unfree Labour* (New York, 1988), pp. 262–79.

———*The Overthrow of Colonial Slavery, 1776–1848* (London, Verso, 1988).

Blake, Stephen P. 'The Patrimonial-Bureaucratic Empire of the Mughals', *Journal of Asian Studies*, 39, 1979, pp. 77–94.

———*Shahjanabad: The Sovereign City in Mughal India, 1639–1739* (Cambridge University Press, 1991).

Bolts, William *Considerations on India Affairs; Particularly Respecting Present State of Bengal and its Dependencies* (3 vols, London, J. Doosley, 1772–75).

Bond, E.A. (ed.) *Speeches of the Managers and Counsel in the Trial of Warren Hastings* (3 vols, London, Longman, Green and others, 1860).

Borthwick, Meredith *The Changing Role of Women in Bengal, 1849–1905* (Princeton University Press, 1984).

Bose, Basanta Coomar *Hindu Customs in Bengal* (Calcutta, Book Company Ltd., 1929).

Boswell C.E., J. Schacht and C. Pellat (eds) *Encyclopedia of Islam,* (vols 1–5, Leiden, E.J. Brill, 1965–86)

Boswell, John *The Kindness of Strangers: The Abandonment of Children in Western Europe from Late Antiquity to the Renaissance* (New York, Pantheon Press, 1988).

Bradford, Helen 'Women, Gender and Colonialism: Rethinking the History of the British Cape Colony and its Frontier Zones, c. 1806–70' in *Journal of African History,* 37, 1996, pp. 351–70.

Bradley, K.R. *Slaves and Masters in the Roman Empire: A Study in Social Control* (Oxford, Oxford University Press, 1987).

Buckler, F.W. 'The Human Khilat', *The Near East and India*, 34, 903, 6 September 1928, pp. 411–22.

Buckley, R.N. *Slaves in Red Coats : The British West India Regiments, 1795–1815* (New Haven, Yale University Press, 1979).

Burnham, Margaret A. 'An Impossible Marriage: Slave Law and Family Law', *Law and Inequality*, 5, 1987, pp. 187–225.

Bush, Barbara *Slave Women in Caribbean Society 1650–1838* (London and Bloomington, Indiana University Press, 1990).

Bush-Slimani, Barbara 'Hard Labour: Women, Childbirth and Resistance in British Caribbean Slave Societies' *History Workshop Journal*, 36, 1993, pp. 83–99.

Butalia, Urvashi 'Community, State and Gender: On Women's Agency During Partition', *Economic and Political Weekly*, 28, 27, 1993, pp. WS12–WS24.

Carre, Abbe *Travels in India and the Near East 1672 to 1674* Lady and Sir Charles Fawcett (trans. and eds) (2 vols, reprint Delhi, Asian Educational Services, 1990).

Caplan, Lionel 'Power and Status in South Asian Slavery' in James L. Watson (ed.), *Asian and African Systems of Slavery* (Oxford, Oxford University Press, 1980).

Carter, Marina 'Indian Slaves in Mauritius (1729–1834)', *Indian Historical Review*, 15, 1987–8, pp. 233–47.

Cassels, Nancy Gardner 'Social Legislation under the Company Raj: the Abolition of Slavery Act V 1843, *South Asia Research*, 11, 1, 1988, pp. 59–88.

Chakrabarti, Sambuddha *Andare Antare: Unish Shatake Bangali Bhadramahila* (Calcutta, Stree, 1995).

Chatterjee, I. 'Testing the Local against the Colonial,' *History Workshop Journal*, 44, 1997, pp. 215–24.

———'Colouring Subalternity: Slaves, Concubines and Social Orphans in Early Colonial India', *Subaltern Studies, X* (Delhi, Oxford University Press, forthcoming).

Chatterji, A. *Bengal in the reign of Aurangzeb* (Calcutta, 1967).

Chatterji, Nandalal 'Shamsuddaulah's Intrigues Against the English', *Bengal Past and Present*, 53, 1, 1937, pp. 31–4

Chattopadhyaya, Amal K. *Slavery in Bengal Presidency 1772–1843* (London, Golden Eagle, 1977).

Chaudhuri, Nanigopal *Cartier, Governor of Bengal, 1769–1772* (Calcutta Firma KLM, 1970).

Chaudhury, Sushil *From Prosperity to Decline: Eighteenth Century Bengal* (Delhi, Manohar, 1995).

Chevers, Norman *A Manual of Medical Jurisprudence for Bengal and North-Western Provinces* (Calcutta, 1856).

Chiba, M. (ed.) *Asian Indigenous Law: In Interaction with Received Law* (London, KPI, 1986).

Chowdhry, Prem *The Veiled Women: Shifting Gender Equations in Rural Haryana 1880–1990* (Delhi, Oxford University Press, 1994).

Clarence-Smith, W.G. (ed.) *The Economics of the Indian Ocean Slave Trade in the Nineteenth Century* (London, Frank Cass, 1989).

Cohn, Bernard S. *An Anthropologist Among the Historians and Other Essays* (Delhi, Oxford University Press, 1990).

Colebrooke, H.T. *A Digest of Hindu Law on Contracts and Succession; with a commentary by Jagannath Tercapanchana: trans. from the original Sanscrit by H.T. Colebrooke* (2 vols, Calcutta, 1802).

Colebrooke, T.E. *The Life of Henry Thomas Colebrooke* (London, 1873).

Coontz, Stephanie *The Social Origins of Private Life: A History of American Families 1600–1900* (Verso, London and New York 1991).

Cooper, Frederick 'The Problem of Slavery in African Studies', *Journal of African History*, 20, 1, 1979, pp. 103–25.

Copeland, Ian *The British Raj and the Indian Princes* (Bombay, Orient Longman, 1982).

Coulson, Noel J. *Conflicts and Tensions in Islamic Jurisprudence* (Chicago, University of Chicago Press, 1969).

Craton, Michael. *Sinews of Empire: A Short History of British Slavery* (New York, Anchor Books, 1974).

Crawford, D.G. 'Letters from Calcutta and Jessore', *Bengal Past and Present*, 4, 1909, pp. 461–85.

Crone, Patricia *Roman, Provincial and Islamic Law: The Origins of the Islamic Patronate* (London and New York, Cambridge University Press, 1987).

Dandamaev, Muhammad A. *Slavery in Babylonia, from Nabopolassar to Alexander the Great (626–331 B.C.)*, V. Powell (trans.) (Illinois, North Illinois University, 1984).

Das, Avinashchandra *Nahar Vamsa Vrittanta* (Calcutta, 1895).

Das, S. *Selections from the Indian Journals: Calcutta Journal* (2 vols, Calcutta, Firma KLM, 1963–5).

Das, Shyamal 'The Mother of Jahangir', *Journal of the Asiatic Society of Bengal*, 57, 1, 1888, pp. 71–5.

Dasgupta, Jotindra Nath: *Bengal in the Sixteenth Century A.D.* (Calcutta, 1914, reprint Delhi, Vintage Books, 1989).

Datta, Kalikinkar *Alivardi and His Times* (Calcutta, World Press, 1963).

Datta, Kalyani *Pinjare Boshiya* (Calcutta, Stree, 1996).

Datta, Pradip K. '"Dying Hindus": the Production of Hindu Communal Common Sense in Early 20th Century Bengal', *Economic and Political Weekly*, 28, 25, 1993, pp. 1305–19.

Davidoff, Leonore and Catherine Hall, *Family Fortunes: Men and Women of the English Middle-class 1780–1850* (London and Chicago, Hutchinson and Chicago University Press, 1987).

Davis, David Brion *Slavery and Human Progress* (New York, Oxford University Press, 1984).

———*The Problem of Slavery in Western Culture* (Ithaca, Cornell Univ. Press, 1966 London.)

Day, Tony 'Ties that (Un) bind: Families and states in Premodern Southeast Asia', *Journal of Asian Studies*, 55, 2, 1996, pp. 384–409.

Day, U.N. *The Mughal Government 1556–1707* (New Delhi, Munshiram Manoharlal, 1969).

Dingwaney, M. and U. Patnaik (eds) *Chains of Servitude* (Madras, Sangam and Oxford University Press, 1985).

Dhagamwar, Vasudha, *Law, Power and Justice: The Protection of Personal Rights in The Indian Penal Code* (New Delhi, Sage Publications, 1992).

Dharampal-Frick, Gita 'Shifting Categories in the Discourse on Caste: Some Historical Observations' in Vasudha Dalmia and Heinrich von Stietencron, *Representing Hinduism: The Construction of Religious Traditions and National Identity* (New Delhi and London, Sage Publications, 1995).

Drescher, Seymour, *Capitalism and Antislavery* (London, Macmillan, 1986).

Dunbar, Janet (ed.) *Tigers, Durbars and Kings: Fanny Eden's Indian Journals 1837–1838* (London, John Murray, 1988).

Dutt, L.P. 'Administration of Justice in Bengal in the Last Decades of the Eighteenth Century', *Bengal Past and Present,* 54, 1–2, 1937, pp. 18–24.

Eaton, Richard M. *The Rise of Islam and the Bengal Frontier 1204–1760* (Berkeley, California University Press, 1993).

Elliott, H.M. *The History of India as Told by its Own Historians* J. Dowson (ed.) (22 vols, London, 1877).

Eltis, David and David Richardson, 'West Africa and the Transatlantic Slave Trade: New Evidence of Long-Run Trends', *Slavery and Abolition*, 18, 1, 1997, pp. 16–35.

Eltis, D. and Walvin, J. (eds) *The Abolition of the Atlantic Slave Trade: Origins and Effects in Europe, Africa, and the Americas* (Madison, University of Wisconsin Press, 1981).

Fawcett, C. *The First Century of British Justice in India: An Account of the Court of Judicature at Bombay established 1672 and of other Courts of Judicature at Bombay, Calcutta and Bombay from 1661 to latter Part of the Eighteenth Century* (Oxford, Clarendon Press, 1934).

———(ed.) *The English Factories of India 1670–1677* (Oxford, Clarendon Press, 1954).

Feierman, Steven 'Africa in History: The End of Universal Narratives', in Gyan Prakash (ed.), *After Colonialism: Imperial Histories and Post Colonial Displacements* (Princeton, Princeton University Press, 1995), pp. 40–65.

Findly, Elison Banks *Nur Jahan: Empress of Mughal India* (Delhi, Oxford University Press, 1993).

Finley, M.I. *Ancient Slavery and Modern Ideology* (London, Chatto and Windus, 1980).

———*The Ancient Economy* (first edition 1973, revised edition Hammondsworth, Penguin, 1992).

Fisch, J. *Cheap Lives and Dear Limbs: The British Transformation of the Bengal Criminal Law 1769–1817* (Weisbaden, 1983).

Fisher, Alan 'Studies in Ottoman Slavery and Slave Trade, II: Manumission', *Journal of Turkish Studies*, 4, 1980, pp. 49–56.

Fisher, A. and G.B. *Slavery and Muslim Society in Africa: The Institution in Saharan and Sudanic Africa and the Trans-Saharan Trade* (London, C. Hurst and Co., 1970).

Fisher, Michael H. *A Clash of Cultures: Awadh, the British and the Mughals* (Delhi, Manohar, 1987).

———*Indirect Rule in India: Residents and the Residency System*, 1757–1857 (Delhi, Oxford University Press, 1991).

———(ed.) *The Politics of the British Annexation of India 1757–1857* (Delhi, Oxford University Press, 1993).

———'Political Marriage Alliances at the Shi 'i Court of Awadh' *Comparative Studies in Society and History,* 25, 4, 1984, pp. 593–616.

Flandrin, Jean Louis *Families in Former Times* Richard Southern (trans.) (Cambridge, Cambridge University Press, 1979).

Fogel, Robert W. and Stanley L. Engerman *'Time on the Cross: the Economics of Negro Slavery,* (2 vols, Boston, Little, Brown, 1974).

Forbes, James *Oriental Memoirs: Selected and Abridged From a Series of Familiar Letters Written During Seventeen Years Residence in India* (4 vols, London, White, Cochrane and Co., 1813).

Forrest, G.W. *The Administration of Warren Hastings, 1772–1785* (Calcutta 1892, reprint Delhi, Inter-India Publications, 1984).

Foster, George *Sketches of the Mythology and Customs of the Hindoos* (London, 1785).

Foster W. (ed.), *The Journal of Sir Thomas Roe, 1615–1619* (London, Oxford University Press, 1926).

Fruzetti, Lina M. *The Gift of a Virgin: Women, Marriage and Ritual in a Bengali Society* (Delhi, Oxford University Press, 1990).

Fukazawa, Hiroshi *The Medieval Deccan: Peasants, Social Systems and States, Sixteenth to Eighteenth Centuries* (Delhi, Oxford University Press, 1991).

Ganguly, Narendranath 'A Peep into the Social Life of Bengal in the 18th Century, *Bengal Past and Present*, 69, 1950, pp. 46–50.

Gardner, Jane F. *Women in Roman Law and Society* (London, Routledge, 1986).

Garlan, Y. *Slavery in Ancient Greece*, Janet Lloyd (trans.) (Ithaca, Cornell University Press, 1982).

Geggus, David Patrick *Slavery, War and Revolution: The British Occupation of St. Domingue, 1793–98* (Oxford, Clarendon Press, 1982).

Genovese, Eugene D. and Elizabeth Fox-Genovese *Fruits of Merchant Capital: Slavery and Bourgeois Property in the Rise and Expansion of Capitalism* (New York, Oxford University Press, 1983).

Ghose, L.N. *A Modern History of the Indian Chiefs, Rajas and Zamindars.* (2 vols, Calcutta, JN Ghosh, 1879–81).

Ghosh, Amitava 'The Slave of Ms. H.6' in Partha Chatterjee and Gyanendra

Pandey (eds), *Subaltern Studies*, VII, (Delhi, Oxford University Press 1993), pp. 159–220.

Ghosh, Benoy *Kolkata Shaharer Itibritta* (Calcutta, Bak-Sahitya, 1975)

Ghoshal, Panchanana *Aparadha Vijnana* (5 vols, Calcutta, Gurudas Chattopadhyay and Sons, 1954).

Gladwin, F. *A Dictionary of Mohammedan Law and of Bengal Revenue Terms With a Vocabulary, Persian and English* (Calcutta, 1797).

Glassman, Jonathon 'No Words of Their Own', *Slavery and Abolition*, 16, 1, 1995, pp. 131–45.

———'The Bondsman's New Clothes: The Contradictory Consciousness of Slave Resistance on the Swahili Coast', *International Journal of African History*, 32, 1991, pp. 277–312.

Gleig, G.R., *Memoirs of the Life of the Right Honorable Warren Hastings* (2 vols, London, Richard Bentley, 1841).

Godse, Vishnubhat *Majha Pravasa*, D.V. Potdar (ed.), (reprint Pune, 1974).

Goode, S.W. *Municipal Calcutta: Its Institutions in Their Origin and Growth* (Edinburgh, 1916).

Goody, Jack *The Oriental, the Ancient and the Primitive* (Cambridge University Press, 1990).

Gordon, M. *Slavery in the Arab World* (New York, New Amsterdam Books, 1989).

Goveia, E.V. *The West Indian Slave Laws of the Eighteenth Century* (Barbados, Caribbean University Press, 1970).

Grand, George F. *The Narrative of the Life of a Gentleman Long Resident in India* (Cape of Good Hope, 1814, reprint Calcutta, Calcutta Historical Society 1910).

Griffin, Lepel H., *The Laws of Inheritance to Chiefships as Observed by the Sikhs Previous to the Annexation of the Punjab* (Lahore, Lahore Book House, 1869).

Guha, Amalendu 'The Medieval Economy of Assam' in T. Raychandhuri and I. Habib (eds) *The Cambridge Economic History of India*, I.C. 1200–1750 (Hyderabad and Cambridge, Orient Longman and Cambridge University Press, 1984).

Guha, Sumit 'An Indian Penal Regime: Maharashtra in the Eighteenth Century', *Past and Present*, 147, 1995, pp. 101–26.

Gupta, Aniruddha, 'Suppression of Slave-Trade and British Imperialist Strategies in the Indian Ocean 1815–1870' in Bissoondoyal and Serwansingh (eds) *Slavery in South West Indian Ocean* (Mauritius, MGI, 1989).

Gutman, H.G. *The Black Family in Slavery and Freedom, 1750–1925* (New York, Vintage Books, 1976).

Guyer, Jane I., 'Wealth in People, Wealth in Things—Introduction', *Journal of African History*, 36, 1995, pp. 83–90.

Habib, Irfan 'Non-Agrarian Production and Urban Economy' in T. Raychaudhuri and I. Habib (eds) *Cambridge Economic History of India* (Hyderabad, Orient Longman and Cambridge, Cambridge University Press, 1984), I.

———'Slavery in the Delhi Sultanate, Thirteenth and Fourteenth Centuries: Evidence from Sufic Literature', *Indian Historical Review*, 15, 1–2, 1988–9, pp. 248–56.

———'Akbar's Social Views: A Study of their Evolution', Papers of the Indian History Congress, 1992–3.

Halhed, N.B. *A Code of Gentoo Laws, Or Ordinations of the Pundits, from a Persian Translation, made from the Original Written in the Shanscrit Language* (London, 1776).

Hall, Catherine, *White, Male and Middle Class: Explorations in Feminism and History* (Cambridge, Polity Press, 1992).

Hambly, Gavin 'Who were the *Chihilgani*, the Forty Slaves of Sultan Shams Al-din Iltutmish of Delhi?', *Iran*, 10, 1972.

———'A Note on the Trade in Eunuchs in Mughal Bengal', *Journal of the American Oriental Society*, 94, 1, 1974, pp. 125–30.

Hamilton, Charles *The Hedaya or Guide : A Commentary on the Mussulman Laws* (4 vols, London, 1791).

Hamilton, Walter *A Geographical, Statistical, and Historical Description of Hindostan and the Adjacent Countries* (2 vols, London, John Murray, 1820).

Haq, S. Moinul *Khafi Khan's History of Alamgir* (English translation of relevant portions of *Muntakhab al- lubab*, Karachi, Pakistan Historical Society, 1975).

Harington, J.H. *Analysis of the Laws and Regulations of the East India Company in Bengal* (3 vols, Calcutta, 1820–21).

Harris, J.E. *The African Presence in Asia : Consequences of the East African Slave Trade* (Evanstone, Northwestern University Press, 1971).

Hasan, Parveen 'Art and Architecture' in Sirajul Islam (ed.) *History of Bangladesh* (3 vols, Dhaka, Asiatic Society of Dhaka Press, 1992).

Hastings, Francis Rawdon *The Private Journal of the Marquess of Hastings* Marchioness of Bute (ed.) (2 vols, London, Saunders and Otley, 1858).

Haynes, D. and G. Prakash (eds) *Contesting Power: Resistance and Everyday Social Relations in South Asia* (Delhi, Oxford University Press, 1991).

Heber, Reginald Rev. *Narrative of a Journey Through the Upper Provinces of India, From Calcutta to Bombay, 1824–25* edited by his wife (2 vols, London, John Murray, 1828).

Hellie, Richard *Slavery in Russia 1450–1725* (Chicago and London, University of Chicago Press, 1982).

Hickey, William *Memoirs of William Hickey* A. Spencer (ed.) (4 vols, London, 1925).

Higman, B.W. *Slave Population and Economy in Jamaica, 1807–1834*, (Cambridge, Cambridge University Press, 1976).

Hiralal, 'Sagar Ka Bundeli Shilalekh', *Nagaripracharini Patrika*, 8, Samvat 1984/1926 C.E., pp. 395–400.

Hopkins, Keith *Conquerors and Slaves* (Cambridge, Cambridge University Press, 1978).

Hunter, W.W. *Bengal Ms. Records* (London, W.H. Allen, 1894).

Hyam, Ronald *Empire and Sexuality* (Manchester University Press, 1990).

Inalcik, Halil *The Ottoman Empire, the Classical Age 1300–1600* N. Itzkowitz and C. Ember (trans.) (London, Weidenfeld and Nicolson, 1973).

Inden, Ronald and Ralph Nicholas *Kinship in Bengali Culture* (Chicago, 1977).

Inden, Ronald *Imagining India* (Oxford, Blackwell, 1990)

Irvine, W. *The Army of the Indian Moghuls: Its Organisation and Administration* (London, 1903, reprint New Delhi, 1962).

———'Bangash Nawabs of Farrukhabad', *Journal of Asiatic Society of Bengal*, 47, 1878, pp. 279–345.

———*Later Mughals*, ed. and augmented J.N. Sarkar (2 vols, reprint Oriental Books, Delhi, 1971).

Ishwardas Nagar *Futuhat - i - Alamgiri* Tasneem Ahmad (trans. and ed.) (Idarah-i-Adabiyat-i Dilli, Delhi, 1978).

Islam, Rafiqul, *Dhakar Katha 1610–1910* (Dhaka, Ahmed Publishing House, 1982)

Islam, Sirajul *The Permanent Settlement in Bengal: A Study of Its Operation, 1790–1819* (Dacca, Bangla Academy, 1979).

———(ed.) *History of Bangladesh 1704–1971* (3 vols, Dhaka, Asiatic Society of Dhaka, 1992).

Jaffur Shurreef *Qanoon-i-Islam* G.A. Herklots (trans.), (London 1832 and revised and ed. W. Crooke Calcutta, 1921).

Jana, Nareshchandra and Manu (eds) *Atmakatha* (2 vols, Calcutta, 1982).

Jaschok, M. *Concubines and Bondservants: The Social History of a Chinese Custom* (London, Zed Books, 1988).

Jaschok, M. and S. Miers *Women and Chinese Patriarchy: Submission, Servitude and Escape* (Hong Kong and London, Hong Kong University Press and Zed Books, 1994).

Jayaswal, K.P. 'A Judgement of a Hindu Court in Sanskrit, *Journal of the Bihar and Orissa Research Society*, 6, 1920, pp. 246–58.

Joseph, Sebastian 'Slave Labour of Malabar in the Colonial Context' in S. Bhattacharya (ed.), *Essays in Modern Indian Economic History* (Delhi, Munshi Manoharlal, 1987), pp. 46–54.

Joshi, Varsha *Polygamy and Purdah: Women and Society among Rajputs* (Jaipur and New Delhi, Rawat Publications, 1995).

Karim, Abdul *Dacca: The Mughal Capital* (Dacca, Asiatic Society of Pakistan, 1964).

Karlekar, Malavika *Voices from Within: Early Personal Narratives* (Delhi, Oxford University Press, 1991).

Kaw, Mushtaq Ahmed 'Some Aspects of *Begar* in Kashmir in the Sixteenth to Eighteenth Centuries', *Indian Economic and Social History Review*, 27, 4, 1990, pp. 465–74.

Khan, Saiyid Ghulam Hussain *Seir Mutaqherin; or View of Modern Times, Being an History of India from the year 1118 to the year 1194.* Haji Mustafa (trans.) (reprint Lahore, 1930).

Khan, Abdul Majed *The Transition in Bengal, 1756–1775: A Study of Saiyid Muhammed Reza Khan* (Cambridge, Cambridge University Press, 1969).

Khan, Shafaat Ahmed (ed.) *John Marshall in India : Notes and Observations in Bengal 1668–1672* (London, Oxford University Press, 1927).

Kidwai, Salim 'Sultans, Eunuchs and Domestics: New Forms of Bondage in Medieval India' in M. Dingwaney and U. Patnaik (eds), *Chains of Servitude* (Madras, Sangam and Oxford University Press, 1985), pp. 76–96.

Kirschenbaum, A. *Sons, Slaves and Freedmen in Roman Commerce* (Jerusalem, Magnes Press, 1987).

Klein, M.A. and Robertson, C.C. *Women and Slavery in Africa* (Madison, Wisconsin University Press, 1983).

Klein, M. A. (ed.) *Breaking the Chains; Slavery, Bondage, and Emancipation in Modern Africa and Asia* (Madison, University of Wisconsin Press, 1993).

Knox, G.E. *The Criminal Law of the Bengal Presidency* (2 vols, Calcutta, Thacker Spink and Co. 1873).

Kolchin, Peter 'Some Recent Works on Slavery, outside the United States: an American Perspective,' *Comparative Studies in Society and History*, 28, 4, 1986, pp. 767–77.

Kolff, Dirk H.A. *Nauker, Rajput and Sepoy: The Ethnohistory of the Military Labour Market in Hindustan, 1450–1850* (Cambridge, Cambridge University Press, 1990).

Kopytoff, I. 'The Cultural Context of African Abolition' in R. Roberts and S. Miers (eds), *The End of Slavery in Africa* (Madison, University of Wisconsin Press, 1988), pp. 485–503.

Kozlowski, Gregory C. *Muslim Endowments and Society in British India* (Cambridge, Cambridge University Press, 1985).

Kuhn, Franz Gottlieb 'The Jat Dynasty of Bharatpur' Jadunath Sarkar (trans.), *Bengal Past and Present*, 75, 1956, pp. 71–86

Kumar, Dharma *Land and Caste in South India* (Cambridge, Cambridge University Press, 1965).

———'Colonialism, Bondage and Caste in British India' in Klein (ed.), *Breaking the Chains* (Madison, 1993), pp. 112–30.

———'Caste and Landlessness in South India' in Gyan Prakash (ed.), *The World of the Rural Labourer in Colonial India* (Delhi, Oxford University Press, 1994), pp. 75–108.

———*Colonialism, Property and the State* (Delhi, Oxford University Press, 1998).

Kumar, Sunil 'When Slaves Were Nobles: The Shamsi *Bandagan* in the Early Delhi Sultanate', *Studies in History*, 10, 1, n.s. 1994, pp. 23–52.

Lal, K.S. *The Mughal Harem* (New Delhi, Aditya Prakashan, 1988).

———*Muslim Slave System in Medieval India* (New Delhi, Aditya Prakashan, 1994).

Lariviere, Richard W. (ed.) *Studies in Dharmasastra* (Calcutta, Firma KLM, 1984).

Laslett, Peter (ed.) *Household and Family in Past Time* (Cambridge, Cambridge University Press, 1972).

Lewin, Capt. T.H. *The Hill Tracts of Chittagong and the Dwellers Therein, with Comparative Vocabularies of the Hill Dialects* (Calcutta, 1869).

Lewis, Bernard *Race and Slavery in the Middle East: An Historial Enquiry* (New York, Oxford University Press, 1990).

Lippman, M., et al. *Islamic Criminal Law and Procedure: An Introduction* (New York, Praeger, 1988).

Llewellyn-Jones, R. *A Very Ingenious Man: Claude Martin in Early Colonial India* (Delhi, Oxford University Press, 1993).

Lovejoy, Paul 'Concubinage and the Status of Women Slaves in Early Colonial Northern Nigeria', *Journal of African History*, 29, 1988, pp. 245–66.

Lovejoy, Paul and Jan S. Hagendorn *Slow Death for Slavery: The Course of Abolition in Northern Nigeria, 1897–1936* (Cambridge, Cambridge University Press, 1993).

Macnaghten, W.H. *Principles and Precedents of Moohummudan Law* (Calcutta, 1825) and edited William Sloan (Calcutta, 1860).

Mahtab, P.C. 'The Bengal Nobles: A Status Group, 1911–19', *Bengal Past and Present*, 92, 1973, pp. 23–36.

Majumdar, Niharkana 'Bengal Politics in 1765', *Bengal Past and Present*, 75, 1956, pp. 98–117.

———*Justice and Police in Bengal: A Study of the Nizamut in Decline* (Calcutta, Firma KLM, 1960).

Malcolm, John *A Memoir of Central India, Including Malwa and Adjoining Provinces* (2 vols, London 1823–24, reprint Delhi, Sagar Publications 1970).

Mallat, C. and J. Connors (eds) *Islamic Family Law* (London, Graham and Trotman, 1990).

Mallik, Kumudnath *Nadia Kahini* (first ed. 1317, third ed., Calcutta 1986).

Mandal, Panchanan *Chithipatre Samajchitra* (2 vols, Santiniketan Viswabharati, 1985).

Mannan, Kazi Abdul (ed.) *Mosharraf Rachana Sambhar* (Dacca, Bangla Academy, 1976).

Manrique, F.S. *Travels of Fray Sebastian Manrique* Col. C.E. Luard (trans. and ed.) (Oxford, Hakluyt Society, 1927).

Manucci, Niccolao *Storia do Mogor or Mogul India 1653–1708* W. Irvine (trans.) (4 vols, Reprint Calcutta, Editions Indian, 1966 of first ed. 1907).

Marglin, F. A. *Wives of the God-King : The Rituals of the Devadasis of Puri* (Delhi, Oxford University Press, 1985).

Marmon, Shaun *Eunuchs and Sacred Boundaries in Islamic Society* (New York, Oxford University Press, 1995).

Marshall, Peter J. *The New Cambridge History of India*, vol. II; *Bengal: The British Bridgehead: Eastern India, 1740–1828* (Cambridge, Cambridge University Press, 1988).

Martin R. M. *The History, Topography and Antiquities of Eastern India* (3 vols, London, 1835).

Mayer, Peter 'Inventing Village Tradition: The Late 19th Century Origins of the North Indian "Jajmani" System', *Modern Asian Studies*, 27, 2, May 1993, pp. 357–95.

Mayne, John D. *The Criminal Law of India* (Madras, Higginbotham and Co., 1896).

McLane, John R. *Land and Local Kingship in Eighteenth–Century Bengal* (Cambridge, Cambridge University Press, 1993).

Meillasoux, C. *The Anthropology of Slavery: The Womb of Iron and Gold*, Alide Dasnois (trans.) (London, Athlone Press, 1991).

Mendelsohn, Isaac *Slavery in the Ancient Near East* (New York, Oxford University Press, 1949).

Menon, Ritu 'Recovery, Rupture, Resistance: Indian State and Abduction of Women During Partition', *Economic and Political Weekly*, 28, 27, 1993, pp. WS2–WS11.

Mernissi, Fatima *The Veil and the Male Elite: A Feminist Interpretation of Women's Rights in Islam.* Mary Jo Lakeland (trans.) (New York, Addison Wesley, 1991).

———*Islam and Democracy: Fear of the Modern Worlds*. Mary Jo Lakeland (trans.) (New York, Addison Wesley, 1992).

———*The Forgotten Queens of Islam*. Mary Jo Lakeland (trans.) (Cambridge, Polity Press, 1993).

Metcalf, Barbara D. 'Narrating Lives: A Mughal Empress, A French Nabob, A Nationalist Muslim Intellectual' *Journal of Asian Studies*, 54, 2, 1995, pp. 474–9.

———(trans.) *'Perfecting Women: Maulana Ashraf 'Ali Thanawi's Bihishti Zewar* (Delhi, Oxford University Press, 1992).

Miers, S. and I. Kopytoff (eds) *Slavery in Africa: Historical and Anthropological Perspectives* (Madison, University of Wisconsin Press, 1977).

Miers, S. and R. Roberts (eds) *The End of Slavery in Africa* (Madison, University of Wisconsin Press, 1988).

Mill, James *History of British India* (H.H. Wilson, London 1858).

Miller, Joseph C. *Way of Death: Merchant Capitalism and the Angolan Slave Trade, 1730–1830* (London, James Currey, 1988).

Mitamura, Taisuke *Chinese Eunuchs: The Structure of Intimate Politics* C.A. Pomeroy (trans.) (Tokyo and Vermont, Charles Tuttle, 1970).

Mitra, K.P. 'Matters Criminal (from Behar Records)', *Bengal Past and Present*, 73, 1954, pp. 127–36.

Mitra, Sivaratan (comp.) *Types of Early Bengali Prose* (Calcutta, Calcutta University, 1922).

Moniz, Antonio Fransisco (compiled and ed.), *Noticias e Documentos para a Historia de Damao: Antiga Provincia do Norte* (10 vols, Bastora, Tipografia Rangel, 1900–17).

Montriou, William A. *Some Precedents and Records to Aid Enquiry as to The Hindu Will of Bengal* (Calcutta, D'Rozario and Co., 1870).

Morgan, Philip D. 'The Cultural Implications of the Atlantic Slave Trade: African Regional Origins, American Destinations and New World Developments', *Slavery and Abolition*, 18, 1, 1997, pp. 122–45.

Morgan, W. and A.G. Macpherson *The Indian Penal Code (Act XLV of 1860)* (Calcutta, 1861).

Moulik, Kshitish C. (ed.) *Prachin Purbabanga Gitika* (5 vols, Calcutta, Calcutta Book House 1971).

Mountmorris, George, Viscount Valentia *Voyages and Travels to India, Ceylon, the Red Sea, Abyssinia and Egypt in the years 1802–1806* (2 vols, London, William Miller, 1811).

Mukharji, T.N. *Art-Manufactures of India; Specially Compiled for the Glasgow International Exhibition* (Calcutta, 1888, reprint Delhi, Navrang 1974).

———*A Handbook of Indian Products* (Calcutta, J. Patterson, 1883).

Mukherjee, S.N. *Sir William Jones, a Study in Eighteenth Century British Attitudes to India* (Cambridge, Cambridge University Press, 1968).

———'A Note on Sir William Jones and the Slave Trade in Bengal', *Bengal Past and Present*, 82, 154, 1963, pp. 106–11.

Mukherji, Tarit K. 'Aldermen and Attorneys–Mayor's Court Calcutta', *Indian Historical Quarterly*, 26, 1, 1950, pp. 51–66.

———'Dispute between Court Cutcherry and Mayor's Court Calcutta', *Indian Historical Quarterly*, 27, 1, 1951, pp. 35–43.

Mukhopadhyaya, Rakhaldas *Bardhaman Rajvamsanucharita* (Calcutta, 1914).

Mundy, Martha 'The Family, Inheritance, and Islam: a Re-examination of the Sociology of Faraid Law' in A. Al-Azmeh (ed.) *Islamic Law: Social and Historical Contexts* (London, Routledge, 1988), pp. 1–123.

Mundy, Peter *The Travels of Peter Mundy in Europe and Asia 1608–1667* by R.C. Temple (ed.) (2 vols, London, Hakluyt Society, 1894).

Munro, Innes *A Narrative of the Military Operations on the Coromandel Coast Against the Combined Forces of the French, Dutch and Hyder Ally Cawn* (London, T. Bensley, 1789).

Murthy, S.S. Ramachandra '*Sanis* in Medieval Andhra Temples' in K.V. Ramesh et al (eds) *Indian History and Epigraphy* (Delhi, Agam Kala, 1990), pp. 188–94.

Nainsi *Muhta Nainsiri Khyat* Badriprasad Sakhariya (ed.) (2 vols, Jodhpur, Rajasthan Oriental Research Institute, 1984).

Nair, Janaki 'The Devadasis, Dharma and the State', *Economic and Political Weekly*, 29, 50, 1994, pp. 3157–67.

Nandi, Somendra Chandra *History of the Cossimbazar Raj in the Nineteenth Century* (vol. I, Calcutta, Dev-All, 1986).

Orme, Robert *Orme's Historical Fragments of the Mogul Empire, of the Morattoes, and of the English Concern in Indostan* (1782, Reprint Delhi, 1974).

Oturkar, Rajaram Vinayak *Peshwekalin Samajik va. Arthik Patravyavahara* (Pune, Bharat Itihas Sanshodhak Mandal, 1950).

Pal, Dharam 'The Influence of the Slaves in the Muslim Administration of India', *Islamic Culture*, 18, 1944 (Reprint London, 1971).

Pandia, Mohanlal Vishnulal 'The Antiquity of Poet Nagari Das and His Concubine Rasik Bihari alias Bani Thani', *Journal of the Asiatic Society of Bengal*, 66, 1, 1897, pp. 63–75.

Panda, Chitta *The Decline of the Bengal Zamindars: Midnapore 1870–1920* (Delhi, Oxford University Press, 1996).

Parks, Fanny *Wanderings of a Pilgrim in Search of the Picturesque During Four-and-Twenty Years in the East; with Revelations of Life in the Zenana* (2 vols, first published 1850 reprint Karachi, Oxford University Press, 1975).

Pateman, Carole *The Sexual Contract* (London, Polity Press, 1988).

Paton, Diana 'Decency, Dependence and the Lash: Gender and the British Debate over Slave Emancipation, 1830–34,' *Slavery and Abolition,* 17, 3, 1996, pp. 163–84.

Patterson, Orlando. *Slavery and Social Death* (Massachusetts, Harvard University Press, 1982).

Pearse, Thomas Deane 'Memoir of Colonel Thomas Deane Pearse of Bengal Artillery', *Bengal Past and Present*, 2, 3, 1908, pp. 305–23; 2, 4, 1908, pp. 459–96.

Peirce, Leslie P. *The Imperial Harem: Women and Sovereignty in the Ottoman Empire* (New York, Oxford University Press, 1993).

Penner, P. and R. Dale Maclean (eds) *The Rebel Bureaucrat—Frederick John Shore* (Delhi, 1983).

Petry, Carl F. 'From Slaves to Benefactors: The Habashis of Mamluk Cairo', *Sudanic Africa*, 5, 1994, pp. 57–66.

Pipes, Daniel *Slave Soldiers and Islam, the Genesis of a Military System* (New Haven, Yale University Press, 1981).

———'Mawlas: Freed Slaves and Converts in Early Islam' in John R. Willis (ed) *Slaves and Slavery in Muslim Africa,* vol. 1 (London, 1985).

Piot, Charles 'Of Slaves and the Gift: Kabre Sale of Kin During the Era of the Slave Trade', *Journal of African History*, 37, 1, 1996, pp. 31–49.

Pomeroy, Sarah B. *Goddesses, Whores, Wives, and Slaves: Women in Classical Antiquity* (London, Pimlico, 1975).

Prakash, Gyan *Bonded Histories: Genealogies of Labour Servitude in Colonial India* (Cambridge, Cambridge University Press, 1990).

———'Terms of Servitude: The Colonial Discourse on Slavery and Bondage in India' in Martin A. Klein (ed.) *Breaking the Chains:* Slavery, Bondage, and Emancipation in Modern Africa and Asia. (Madison, University of Wisconsin Press, 1993), pp. 131–49.

———(ed.), *After Colonialism: Imperial Histories and Post-Colonial Displacements* (Princeton, Princeton University Press, 1995).

———(ed.), *The World of the Rural Labourer in Colonial India* (Delhi, Oxford University Press, 1994).

———'Colonialism, Capitalism and the Discourse of Freedom' in Shahid Amin and Marcel van der Linden (eds) *International Review of Social History*, Supplement 4, 1996, pp. 9–25.

Qureshi, I.H. *The Administration of the Mughul Empire* (New Delhi, N.V. Publications 1979).

Rajwade, V.K. (ed.) *Bharata Itihas Sanshodhaka Mandal Varshik Itivritta* (Pune, BISM Press 1923).

Rapp, R., Ross, E. and Bridenthal, R. 'Examining Family History' in J. Newton, J. Ryan and J. Walkowitz (eds) *Sex and Class in Women's History, Essays From Feminist Studies* (London, Routledge, 1983), pp. 232–57.

Rawson, Beryl (ed.) *The Family in Ancient Rome: New Perspectives* (London, Routledge 1986).

Ray, Charu Chandra 'Slaves and Slavery in Old Chandernagore', *Bengal Past and Present*, 6, 1910, pp. 257–65.

Ray, Dinabandhu *Sekaler Smriti* (Calcutta, Ananda Publishers, 1977).

Ray, Karttikeya Chandra *Kshitish Vamsavali Charita* (Calcutta, 1932).

Raychaudhury Tapan *Bengal under Akbar and Jahangir* (Delhi, Munshiram Manoharlal, 1966).

Rege, Sharmila 'The Hegemonic Appropriation of Sexuality : The Case of the *Lavani* Performers of Maharashtra', in Patricia Uberoi (ed.) *Social Reform, Sexuality and the State* (New Delhi, Sage Publications, 1996).

Reid, Anthony (ed.) *Slavery, Bondage and Dependency in South East Asia* (St. Lucia, London and New York, University of Queensland Press, 1983).

Richards, J.F. *The Mughal Empire* (Cambridge, Cambridge University Press, 1992).

———*Power, Administration and Finance in Mughal India* (London, Variorum, 1993).

Robb, Peter (ed.) *Dalit Movements and the Meanings of Labour in India* (Delhi, Oxford University Press, 1993).

(ed.) *The Concept of Race in South Asia* (Delhi, Oxford University Press, 1995).

Robb, Peter et al (eds), *Society and Ideology: Essays in South Asian History* (Delhi, Oxford University Press, 1993).

Robertson, C. and M.A. Klein (eds) *Women and Slavery in Africa* (Madison, Wisconsin University Press, 1983).

Rousseau, Samuel *A Dictionary of Words Used in the East Indies with Full Explanations* (London, J. Lewell 1805).

Roy, Biswanath *A Peep Through the Hazarduary Palace of Murshidabad* (Berhampore, Baimela Publications, 1981).

Roy, Rammohun *A Second Defence of the Monotheistical System of the Veds in Reply to an Apology for the Present State of Hindoo Worship* (Calcutta, 1817).

Sajawandi, Siraj-al din *Al-Sirajiyyah; Or the Mohammedan Law of Inheritance With a Commentary by Sir William Jones* (Calcutta, 1792).

Salim, Ghulam Husain *Riyazu-us-Salatin or a History of Bengal* Abdus Salam, (trans.) (reprint Idarah-i Adabiyat-i Dilli, Delhi, 1975).

Samiruddin, Maulvi *Bedar Ghafelin* (Calcutta, 1877).

Sandeman, H (ed.) *Selections from Calcutta Gazette.* (Calcutta, 1869).

Sangari, K. and Vaid, S. (eds), *Recasting Women: Essays in Colonial History* (New Delhi, Kali, 1989).

Sardesi, G.S. (ed.) *Selections from the Peshwa Daftar* (Bombay, 1931–4)

Sarkar, Jadunath (ed.) *Bengal Nawabs* (Calcutta, Asiatic Society, 1952).

———*The Fall of the Mughal Empire* (Calcutta, M.C. Sarkar, 1950).

Sarkar, Tanika 'Bondage in the Colonial Context' in Utsa Patnaik and Manjari Dingwaney (eds), *Chains of Servitude* (Madras, 1985).

Sastri, Sivnath *Atmacharit* (Calcutta, Signet Press, 1952).

Savage, Elizabeth *The Human Commodity: Perspectives on the Trans-Saharan Slave Trade* (London, Frank Cass, 1992).

Sen, Dinesh Chandra *Eastern Bengal Ballads* Ramtanu Lahiri Fellowship Lectures for 1922–24 (4 vols Calcutta, the University, 1923–1932).

———*Bengali Prose Style: 1800–1857* (Ramtanu Lahiri Fellowship Lectures, 1919–20, Calcutta, the University, 1921).

———*History of Bengali Language and Literature* (Calcutta, the University, 1911.)

———*Glimpses of Bengali Life: Stray Notes on Some Bengali Ballads* (Calcutta, the University, 1925).

Seton-Karr, W.S. *Selections from the Calcutta Gazettes of the years 1784–1799* (2 vols, Calcutta, 1864–5).

Sharp, Granville *The System of Colonial Law compared with the Eternal Laws of God; and with the Indispensable Principles of the English Constitution* (London, 1807).

Shell, Robert C.-H. *Children of Bondage: A Social History of the Slave Society at the Cape of Good Hope, 1652–1838* (Hanover, Wesleyan University Press, 1988).

Shome, J.G. *Law Reporter* (2 vols., Calcutta, 1878).

Shore, F. J. *Memoirs and Life of Lord Teignmouth* (2 vols, London, 1837).

Shore, John (Lord Teignmouth) (ed.) *Collected Works of Sir William Jones* (9 vols, London, 1807).

Shyllon, F.O. *Black Slaves in Britain* (London, Oxford University Press, 1974).

Sikainga, Ahmad A. 'Shari'a Courts and the Manumission of Female Slaves in the Sudan, 1898–1939', *International Journal of African Historical Studies*, 28, 1995, pp. 1–24.

Singha, Radhika 'The Privilege of Taking Life: Some "Anomalies" in the Law of Homicide in the Bengal Presidency', *IESHR*, 30, 2, 1993, pp. 181–214.

———'Civil Authority and Due Process: Colonial Criminal Justice in the Banaras Zamindari, 1781–1795', Paper at Changing Conceptions of Rights and Justice in South Asia, 16–19 March, 1994.

———'"Making the Domestic More Domestic": Criminal Law and the "Head of the Household"', 1772–1843', *IESHR*, 33, 3, 1996, pp. 309–43.

———*A Despotism of Law: Crime and Justice in Early Colonial India* (Delhi, Oxford University Press, 1998).

Sinha, N.K. 'The Case of Mirza Jan Tuppish: A Treason Trial of 1800', *Bengal Past and Present*, 72, 1953, pp. 39–42.

Skipwith, F. *The Magistrate's Guide* (Calcutta, 1843).

Stanhope, Philip D. *Genuine Memoirs of Asiaticus* W. K. Firminger, (ed.) (Calcutta, 1909).

Stavorinus, John Splinter *Voyages to the East Indies* trans. Samuel Hull Wilcocke (2 vols., London, 1798).

Stewart, Charles *The History of Bengal from the First Mohammedan Invasion until the Virtual Conquest of that Country by the English* A.D. *1757* (reprint Delhi, Oriental Publishers, 1971).

Stolcke, Verena 'The Slavery Period and its Influence on Household Structure and the Family: Jamaica, Cuba and Brazil' in Elza Berquo and Peter Xenos (eds), *Family Systems and Cultural Change* (Oxford, Oxford University Press, 1992).

Stone, Lawrence *The Family, Sex and Marriage in England, 1500–1800* (New York, Harper and Row, 1979).

Strange, Thomas *Hindu Law: Principally With Reference to Such Portions of it as Concern the Administration of Justice in the King's Courts in India* (2 vols, London, 1830, reprint Madras, 1859).

Subrahmanyam, Sanjay 'The Mughal State—Structure or Process? Reflections on Recent Western Historiography', *Indian Economic and Social History Review*, 29, 3, 1992, pp. 291–321.

Tavernier, Baptiste *Travels in India* V. Ball (trans.) (2 vols, London, 1889).

Tayler, William *Thirty Eight Years in India* (London, 1881).

Taylor, James *A Sketch of the Topography and Statistics of Dacca* (Calcutta, 1840).

Tehmas Nama, An Autobiography of a Slave abridged and edited by P. Setu Madhava Rao, (Bombay, Popular Prakashan 1967).

Teignmouth, John *Memory of the Life of John, Lord Teignmouth,* C.J.S. Teignmouth (ed.) (2 vols, London, Hatchard, 1843).

Terry, Edward *A Voyage to East India; Wherein Some Things are Taken Notice of* (London, 1777).

Thakur, Upendra 'Some Aspects of Slavery in Mithila in the 17th–19th Centuries', *Journal of Bihar Research Society,* 64, 1 and 2, 1958, pp. 47–52.

Thevenot and Careri *Indian Travels* S.N. Sen (ed.) (Delhi, National Archives of India, 1949).

Thompson, Edward J. *The Life of Charles, Lord Metcalfe* (London, Faber and Faber 1937).

Toledano, Ehud R. *State and Society in Mid-Nineteenth Century Egypt* (Cambridge University Press, 1990).

———'The Imperial Eunuchs of Istanbul: From Africa to the Heart of Islam', *Middle Eastern Studies*, 20, 1984, pp. 379–90.

———'Slave Dealers, Women, Pregnancy, and Abortion: The Story of a Circassian Slave-Girl in Mid-Nineteenth Century Cairo', *Slavery and Abolition*, 2, 1, 1981, pp. 53–68.

Treggiari, Susan *Roman Marriage* (Oxford, Clarendon Press, 1993).

Tucker, Henry Carre *My Notebook of Rules and Regulations, Collectory and Faujdary* (Calcutta, 1850).

Tushnet, Mark V. *The American Law of Slavery: Considerations of Humanity and Interest* (Princeton University Press, 1981).

———'Critical Method in Legal History', *Cardozo Law Review*, 9, 1985, pp. 997–1011.

Uberoi, Patricia (ed.), *Family, Kinship and Marriage in India* (Delhi, Oxford University Press, 1994).

———*Social Reform, Sexuality and the State* (New Delhi, Sage, 1996).

Vad, G.C., K.B. Marathe, D.V. Parasnis and P.B. Joshi: *Selections from the Satara Rajas and Peshwa's Diaries* (9 vols, Pune, Deccan Vernacular, Translation Society, 1902–11)

Vansittart, Henry *A Narrative of the Transactions in Bengal from the Year 1760 to the Year 1764, during the Government of Mr. Henry Vansittart* (2 vols London, J. Newberry, Dodwey etc. 1766).

Verelst, H. *A View of the Rise, Progress and Present State of the English Government in Bengal* (London, J. Nourse, 1766).

Vijaya, T.P. 'Aspects of Slavery in Coorg in the Nineteenth Century', *Indica*, 29, 1992, pp. 107–22.

———'Honour in Chains: The Problem of *hitti-bitti-chakri* in *Jamma* Tenure in Coorg, 1800–1939', *IESHR*, 32, 2, 1995, pp. 135–53.

Wagoner, Philip B. '"Sultan among Hindu Kings": Dress, Titles and the Islamicization of Hindu Culture at Vijayanagara', *Journal of Asian Studies*, 55, 4, 1996, pp. 851–80.

Waheeduzzaman, Abu Mohammed 'Land Resumption in Bengal 1819–1846, Unpublished Ph.D. Diss., University of London, 1969.

Walvin, James *Black Ivory: A History of British Slavery* (Washington, Howard University Press, 1994).

Watson, Alan *Slave Law in the Americas* (New York and London, University of Georgia Press, 1989).

———'On *The American Law of Slavery 1810–1860*', *Yale Law Journal*, 91, 5, 1982, pp. 1034–47.

———'Seventeenth Century Jurists, Roman Law and the Law of Slavery' and 'Thinking Property at Rome' in *Chicago-Kent Law Review*, 68, 1993, pp. 1343–71.

Watson, James L. (ed) *Asian and African Systems of Slavery* (Oxford, Basil Blackwell, 1980).

Wilbur, C. M. *Slavery in China during Former Han Dynasty 206 B.C.–A.D. 25* (Chicago, Chicago University Press, 1943).

Williams, Eric *Capitalism and Slavery* (Chapel Hill, University of North Carolina Press, 1944).

Willis, John R. (ed.) *Slaves and Slavery in Muslim Africa* (2 vols, London, Frank Cass, 1985).

Wilson, C.R. *The Early Annals of the English in Bengal* (3 vols, Calcutta, Thacker Spink, 1900–17).

Wise, J. *Notes on the Races, Castes and Trades of Eastern Bengal* (London, 1883).

Wolf, Arthur P. and Chieh-shan Huang *Marriage and Adoption in China 1845–1945* (California, Stanford University Press, 1980).

Zaretsky, Eli *Capitalism, the Family and Personal Life* (New York, Harper and Row, 1976).

Index

30106021573463

Gender, Slavery and Law in Colonial India